PLASTIC DESIGN

TITLES OF RELATED INTEREST

Boundary element methods in solid mechanics
S. L. Crouch & A. M. Starfield

Computers in construction planning and control
M. J. Jackson

Geomorphological hazards in Los Angeles
R. U. Cooke

Geology for civil engineers
A. C. McLean & C. D. Gribble

Numerical methods in engineering and science
G. de Vahl Davis

Structural dynamics
H. M. Irvine

Theory of vibration with applications
W. Thomson

PLASTIC DESIGN

AN IMPOSED HINGE-ROTATION APPROACH

P. ZEMAN

Consulting Engineer, Sydney, Australia

and

H. M. IRVINE

Department of Structural Engineering, University of New South Wales, Sydney, Australia

London
ALLEN & UNWIN
Boston Sydney

Allen & Unwin (Publishers) Ltd,
40 Museum Street, London WC1A 1LU, UK

Allen & Unwin (Publishers) Ltd,
Park Lane, Hemel Hempstead, Herts HP2 4TE, UK

Allen & Unwin Inc.,
8 Winchester Place, Winchester, Mass. 01890, USA

Allen & Unwin (Australia) Ltd,
8 Napier Street, North Sydney, NSW 2060, Australia

First published in 1986

British Library Cataloguing in Publication Data

Zeman, P.
Plastic design: an imposed hinge rotation approach.
1. Structural design
I. Title II. Irvine, H. Max
624.1'72 TA658
ISBN 0-04-624009-8

Library of Congress Cataloging in Publication Data

Zeman, P. (Peter), 1946-
Plastic design, an imposed hinge-rotation approach.
Bibliography: p.
Includes index.
1. Plastic analysis (Theory of structures)
2. Structural design. I. Irvine, H. Max. II. Title.
TA652.Z46 1986 624.1'7 85-27492
ISBN 0-04-624009-8 (alk. paper)

Set in 10 on 12 point Times by
Mathematical Composition Setters Limited
7 Ivy Street, Salisbury, Wiltshire, England
and printed in Great Britain by
Butler & Tanner Ltd, Frome and London

To Paula and Sue

Preface

Structures are commonly designed on the basis of either elastic or rigid–plastic theories. However, the division is somewhat arbitrary, and it often detracts from the real object of obtaining the best overall solution. What is actually required in most practical applications is a compromise between the two philosophies. Such a compromise can be achieved by using a far more general design method based on the imposed hinge-rotation approach. No prior assumptions are made about either the extent or the locations of yielding in the structure under its imposed loads. The design solution is defined solely by the limit state design criteria which are imposed.

This monograph is an attempt to describe this general method in as simple a manner as possible. It also serves as an introduction to the numerical technique of linear programming. Attention is directed to both analysis and design, and emphasis is placed on the physical significance of the various calculations involved. In particular, linear programming is described not only as a method of optimization, but also as a method of tracing the gradual development of yielding in a structure until failure.

A minimum of technical jargon is used in the discussion of the numerical procedures. The reader need not have any previous knowledge of the subject, but he is expected to be familiar with methods of elastic analysis – particularly the matrix stiffness method. Extensive use of examples has been made in the monograph. One example which receives considerable attention is the two-span continuous beam described in Chapter 2. This example is used to introduce the method and to illustrate its basic concepts, both numerically and graphically. The method is described in a more general form in Chapter 3. Two other examples which are considered are the portal and high-rise frames discussed in Chapter 4. The examples are used to demonstrate the application of the method, using a computer program known as SODA (System for Optimum Design and Analysis).

For convenience of presentation, the monograph has been written on the basis of first-order theory only. Second-order theory, i.e. the consideration of P–Δ effects and buckling in design, will be the subject of a subsequent publication.

P. Zeman
H. M. Irvine

Acknowledgements

We wish to show our appreciation to all those who helped, both in the research of the present work and in its final presentation.

The senior author would like to thank H. B. Harrison and other staff at the University of Sydney for their assistance during his Ph.D studies. He would also like to thank the School of Civil Engineering at the University of New South Wales for the provision of a research and teaching appointment that made this book possible.

We wish to thank Ch. Massonnet, M. R. Horne and A. R. Toakley for their constructive comments and criticisms. We would also like to acknowledge the tireless efforts of Colin Wingrove and Ruth Rogan in the preparation of the manuscript.

Contents

Preface page ix

Acknowledgements xi

List of tables xiv

Notation xv

1 Introduction 1

1.1 Research in plastic design 1
1.1.1 Heyman's approach 1
1.1.2 Imposed hinge-rotation approach 2
1.2 The proposed design approach 3
References 4

2 Continuous beam example 6

2.1 Analysis 7
2.1.1 Elastic analyis 7
2.1.2 Elastic–plastic analysis 9
2.2 Analysis–design analogy 20
2.3 Elastic design 22
2.3.1 Design space 24
2.3.2 Design calculations 27
2.4 Plastic Design 32
2.4.1 Design space 33
2.4.2 Design calculations 38
Reference 44

3 The proposed design approach 45

3.1 Specification 45
3.1.1 Member size variables 46
3.1.2 Geometry and prestress variables 50
3.1.3 Objective function 50
3.1.4 Yield constraints and hinge-rotation variables 51
3.1.5 Additional constraints 52
3.2 Initialization 53
3.3 Formulation of the response equations 53
3.3.1 Calculation of response 54
3.3.2 Response sensitivities 59
3.4 Formulation of the linear program 64
3.4.1 Objective function 64
3.4.2 Design constraints 65
3.5 Optimization 70
3.5.1 Initial solution 70
3.5.2 Load history phase 72
3.5.3 Design search phase 73
References 74

4 Design examples 75
4.1 Low-rise industrial frame 75
4.1.1 Rigid–plastic design 77
4.1.2 Deflection constrained design 78
4.2 Multistorey frame in seismic zone 79
4.2.1 Rigid–plastic design 81
4.2.2 Strong column design 82
References 84

Index 85

List of tables

2.1 Four alternative design solutions *page* 35
4.1 Optimum design solutions 81

Notation

A	statics matrix or cross-sectional area
E	Young's modulus
G	abbreviated form of SA^{T}
I	second moment of area
K	frame stiffness matrix $(=ASA^{T})$
L	length
M	bending moment
M^{el}	elastic bending moment
M_p	plastic moment capacity $(=f_y Z_p)$
N	an arbitrary large number
P	axial force
P_y	axial force capacity $(=f_y A)$
$\mathbf{Q}$	part of stress resultant solution $(=\mathbf{SP}+\mathrm{S}\boldsymbol{\phi})$
R	hinge rotation variable $(=(EI/L)\phi)$
S	member stiffness matrix or slack variable
$\mathbf{SP}$	part of stress resultant solution
$\mathbf{SR}$	stress resultant vector
W	load parameter
W_0	value of W at factored design loads
$\mathbf{W}$	load vector
$\mathbf{X}$	nodal deformation vector
Z_p	plastic section modulus
f_y	yield stress
$\mathbf{g}$	acceleration due to gravity
m	section property $(=M_p/P_y)$
n	stiffness ratio
p_i, p_0	representations of design variables in linear program
r	alternative form for hinge-rotation variable
v, v_i	design variable
w_0	value of W at working load
$\mathbf{x}$	member deformation vector
z	objective function
α	inclination of a member to the horizontal axis
α_i	coefficient of design variable in constraint equation
β_j	coefficient of hinge rotation variable in constraint equation
γ	load factor parameter $(=\lambda/\lambda_0)$
η_i	coefficient of design variable in objective
θ	joint rotation
λ	load factor
λ_0	design load factor
μ, ν	representations of design variables
ρ	density
ϕ	plastic hinge rotation
$\boldsymbol{\phi}$	plastic hinge rotation vector
Δ	deflection
Δ^{el}	elastic deflection

1

Introduction

In the case of structural analysis, application of the computer was met with hesitation at first. Its potential was not fully realized, perhaps because the first programs were based on traditional methods, which were more suited to manual analysis. As computer technology improved, applications to analysis became more frequent, and attention was redirected to methods that were more systematic, such as the classical stiffness approach. Of course, this approach now forms the basis of most of the computer programs so widely used today.

Just as computers have become increasingly influential in analysis, they are also achieving an increasing importance in design. This importance has been accentuated, firstly, by developments in man–machine interaction and, secondly, in their potential for leading to structures that are more economical. More significantly, they also have the potential for improving the efficiency of the design process itself. However, as was previously the case in computer analysis, these potentials were initially not fully realized. There was once again a tendency to automate only the well established and traditional manual methods.

More recently, the introduction of limit state design concepts has encouraged interest in approaches that are more systematic and fundamental. These approaches are ideally suited to computer application, and their main feature is their inherent flexibility. They provide the opportunity for considering strength, ductility and deflection criteria simultaneously, and they bridge the gap between elastic and rigid–plastic design.

1.1 Research in plastic design

These approaches have developed from recent research in the field of plastic design. This research can be divided into two categories, which are briefly outlined below.

1.1.1 Heyman's approach

The first attempt to develop a systematic approach to plastic design was made by Heyman (1951). The approach is based on rigid–plastic theory, and attention is directed to the design of frames in steel. The section sizes of the members are initially unknown, and the design problem is one of determining the sizes which correspond to a structure of minimum weight. In rigid–plastic design there is no need to consider compatibility equations, and the formulation is based on an arbitrarily released structure. In Heyman's original approach, which is concerned with rigid–plastic collapse criteria only, the design problem reduces to one of linear optimization. The problem can therefore be solved in one application of the linear programming algorithm, a well established numerical tool in the field of operations research (Gass 1969).

Extensions to Heyman's design approach have been proposed by Toakley (1969) and Horne and Morris (1973) for the consideration in design of P–Δ (instability or second order theory) effects and working load deflection criteria. Although these extensions represent considerable improvements, they still rely on a formulation based on an arbitrarily released structure. The opportunity for calculating deflections and formulating deflection criteria is consequently lost, and supplementary routines which return attention to the original structure are required. These routines are then either relatively costly in terms of computing time, or approximate and specialized. Furthermore, because the methods are merely extensions to the rigid–plastic design approach, they are subject to many of its limitations. They still rely implicitly on the formation of a rigid–plastic mechanism, and the associated assumption that compatibility conditions can be neglected at ultimate load. As has been clearly demonstrated by Wood (1958), this assumption only applies if the P–Δ effects are negligible. A policy of ignoring compatibility conditions in a design method which incorporates P–Δ effects is therefore not strictly valid. More significantly, no account is taken of the ductility capacity of the structure and its members. It is assumed that, firstly, plastic hinges can occur anywhere in the structure and, secondly, all hinges that form are capable of undergoing an indefinite rotation. A consequence of these assumptions is that these methods have not been completely successful in controlling either general deflection response or plastic hinge development.

1.1.2 Imposed hinge-rotation approach

A completely separate line of research has proved to be much more promising. This research is based on the imposed hinge-rotation concept introduced by Colonnetti (1955) and subsequently developed by Macchi (1966, 1972). The reason that the concept has not been attractive in the past is that it represents a complete departure from established manual methods, and is therefore still relatively unknown, particularly to practising engineers.

Instead of arbitrarily releasing the structure, calculations begin with the structure in its original elastic state. Response is expressed as a function of an elastic contribution and a series of imposed plastic hinge rotations, which are initially represented as variables. Any response can be expressed in this manner, and this includes not only bending moments and axial forces, but also deflections. The consideration of deflections and P–Δ effects can therefore be integrated much more easily into the design procedure. Moreover, since response is expressed in terms of hinge-rotation variables, the extent of yielding in a design can be controlled simply by imposing restrictions on the values of these variables.

The hinge-rotation concept has been applied to the design of both concrete and steel structures. In the case of concrete, research has been directed considerably more towards analysis than towards design. In analysis, a trilinear relationship between bending moment and hinge rotation is assumed, and efforts are concerned mainly with the development of efficient computational algorithms for considering the numerical problems involved (Maier & Munro 1982). In design, member sizes and stiffnesses are assumed to be known, and the design problem reduces to one of determining the most economical reinforcement details (Krishnamoorthy & Munro 1972).

In the case of steel, efforts to apply the concept to design were initially confined to the determination of rigid–plastic solutions (Davies 1972). More recently, Zeman (1975) developed a far more general method which effectively bridges the gap between rigid–plastic and elastic design. His work culminated in a computer program known as SODA (System for Optimum Design and Analysis).

1.2 The proposed design approach

SODA is one of a number of programs which are becoming available for the plastic design of structures. Another, for example, is one developed by Hung (1984) for design in both concrete and steel. The objective of this monograph is to provide the reader with a basis for understanding the function of these programs and the principles on which they are based. Particular attention is placed on the imposed hinge-rotation concept and the numerical tool of linear programming.

For convenience of presentation, attention is directed solely to steel design and the method proposed by Zeman (1975). Using this method, no assumption need be made concerning either the extent or locations of yielding in the final structure. The method is a generalization of the optimum elastic design approach described by Reinschmidt *et al.* (1966). If the designer chooses to prevent plastic deformation completely, then the solution is an elastic one. On the other hand, if no restrictions are placed on plastic deformation, then the calculations lead to a rigid–plastic solution. In most practical applications, the design solution lies between these two extremes.

The design problem is therefore generally non-linear, and is solved iteratively. As in optimum elastic design, an initial estimate of the design solution is chosen, and is successively improved in a series of design cycles until convergence is obtained. Each cycle involves deriving a linear approximation to the design problem, and then determining its solution using the linear programming algorithm. As will be described, the algorithm is used not only in a design role, but also as a means of determining the load history response of the structure.

In general, convergence is obtained relatively rapidly. In particular, if the solution is a rigid–plastic one, then the design problem is linear (as in Heyman's formulation) and only one cycle is required.

Using the proposed method, the design problem is defined in terms of the following characteristics.

(a) *Criteria.* Strength, ductility or deflection requirements, at either working load or ultimate load, for any number of load conditions.
(b) *Variables.* Member section sizes, geometry or initial prestress in the frame.
(c) *Objective.* Unless otherwise stated, the design objective is to minimize the frame weight.

References

Colonnetti, G. 1955. *L'équilibre des corps déformables.* Paris: Dunod.

Davies, J. M. 1972. A new formulation of the plastic design problem for plane frames. *Int. J. Num. Methods Engng* **5**, 185–92.

Gass, S. I. 1969. *Linear programming.* New York: McGraw-Hill.

Heyman, J. 1951. Plastic design of beams and frames for minimum material consumption. *Q. Appl. Math.* **8**, 373–81.

Horne, M. R. and L. J. Morris 1973. The optimum design of multi-storey rigid frames. In *Optimum structural design – theory and applications*, R. H. Gallagher and O. C. Zienkiewicz (eds), 267–82. New York: Wiley.

Hung, N. D. 1984. CEPAO – an automatic program for rigid–plastic and elastic plastic analysis and optimization of frame structures. *Engng Struct.* **6**, 33–51.

Krishnamoorthy, C. S. and J. Munro 1972. *Optimated design of reinforced concrete frames using linear programming.* Int. Symp. on computer-aided structural design, University of Warwick, UK.

Macchi, G. 1966. *Méthode des rotations imposées.* Structures hyperstatiques (Projet d'annexe aux Recommandations Practiques) – Comité Européen du Béton.

Macchi, G. 1972. *Non-linear analysis and limit design.* Theme report, Int. Conf. on Planning and Design of Tall Buildings, Vol. III, Lehigh University, Pennsylvania, 1–22.

REFERENCES

Maier, G. and J. Munro 1982. Mathematical programming applications to engineering plastic analysis. *Appl. Mech. Revs* **35** (12), 1631–43.

Reinschmidt, K. F., C. A. Cornell and J. F. Brotchie 1966. Iterative design and structural optimization. *J. Struct. Divn, ASCE* **89**(ST6), 281–318.

Toakley, A. R. 1969. The optimum plastic design of unbraced frames. *Civ. Engng Trans, Instn Engrs, Austral.* **CE11**(2), 111–6.

Wood, R. H. 1958. The stability of tall buildings. *Proc. Instn Civ. Engrs* **11**, 69–102.

Zeman, P. 1975. *Optimum elastic–plastic design of framed structures, Vols I and II.* Ph.D. thesis, University of Sydney, Australia.

2

Continuous beam example

Before describing the general design approach, it is convenient to confine attention to a specific example; the two-span continuous beam shown in Figure 2.1. As shown in the figure, two distinct cross sections in the beam can be identified, and the corresponding plastic moment capacities are M_p and M'_p. The beam is assumed to be subjected to a single load condition, and the loads are expressed in terms of the load parameter, W. The factored loads that the beam is required to sustain safely without violating any design criteria are defined by $W = W_0$.

Attention is initially directed to the analysis of the beam, and then to its design. As will be seen, there is an analogy between the two, and this will be discussed.

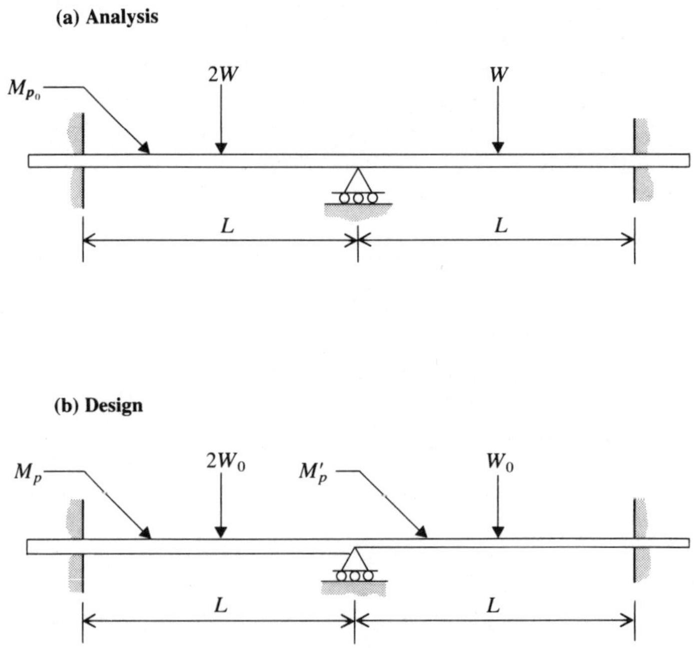

Figure 2.1 Continuous beam example.

2.1 Analysis

In the analysis of the beam, as indicated in Figure 2.1a, the sizes of the cross sections are specified beforehand. For the purposes of this example, these sizes are assumed to correspond to a beam of uniform cross section. For ease of presentation, they are expressed in terms of the design load level, W_0, as follows:

$$M_{p_0} = M'_{p_0} = 0.5W_0L \tag{2.1}$$

$$EI_0/L = EI'_0/L = 3.6W_0L \tag{2.2}$$

where M_{p_0} and I_0 are the moment capacity and second moment of area of the cross section in the first span, M'_{p_0} and I'_0 denote the corresponding values in the second span, E is the Young's modulus and W_0 represents the load level for which the beam is to be designed.

In the following discussion, and also throughout this monograph, yielding is assumed to be characterized by a unit shape factor. The onset of yielding at a cross section therefore coincides with its full plastification as soon as the corresponding bending moment reaches its limiting value, M_p. This bending moment then remains at its limiting value while the cross section undergoes permanent deformation.

The analysis of the beam involves determination of its response for gradually increasing values of the load parameter, W. The analysis is a systematic one, and each value chosen for W corresponds to a distinct stage in the development of progressive yielding. The procedure is continued until the maximum value of W is reached, and the beam collapses. This maximum value is referred to as the load capacity of the beam. If it is found to exceed the specified value W_0, then the moment capacities defined by Equation (2.1) represent a safe design.

2.1.1 *Elastic analysis*

An initial estimate of the maximum value of W is the value required to cause first yielding in the beam. As will now be described, this value can be determined by carrying out an elastic analysis.

Referring to Figure 2.2, the bending moments in the uniform beam can be expressed in terms of slope deflection equations, which are as follows:

$$\begin{aligned} M_{AC} &= \frac{2EI_0}{L}\theta_C - \frac{WL}{4} & M_{CA} &= \frac{4EI_0}{L}\theta_C + \frac{WL}{4} \\ M_{CE} &= \frac{4EI_0}{L}\theta_C - \frac{WL}{8} & M_{EC} &= \frac{2EI_0}{L}\theta_C + \frac{WL}{8} \end{aligned} \tag{2.3}$$

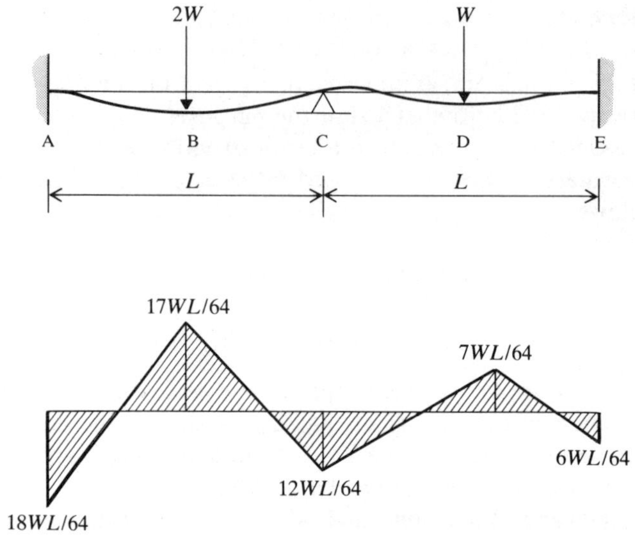

Figure 2.2 Elastic bending moment distribution.

where θ_C is the rotation at joint C. The sign convention is such that clockwise bending moments and rotations are positive. From joint equilibrium at C,

$$M_{CA} + M_{CE} = 0$$

Therefore,

$$\frac{8EI_0}{L}\theta_C + \frac{WL}{4} - \frac{WL}{8} = 0$$

i.e.

$$\theta_C = -WL^2/64EI_0 \tag{2.4}$$

Substituting Equation (2.4) into Equations (2.3), the bending moments are

$$M_{AC} = -18WL/64$$

$$M_{CA} = -M_{CE} = 12WL/64$$

$$M_{EC} = 6WL/64$$

The bending moments at other points along the beam can be deduced from the bending moment diagram shown in Figure 2.2.

Deflections can be determined using the moment-area method. Referring to Figure 2.2, the deflection at B, for example, is Δ_B, where

$$\Delta_B = \theta_A L/2 - (2M_A + M_B)\frac{L^2}{24EI_0}$$

$$= 0 - \left[\left(-2 \times 18\frac{WL}{64}\right) + 17\frac{WL}{64}\right]\frac{L^2}{24EI_0}$$

$$= \frac{19WL^3}{1536EI_0}$$

i.e.

$$\frac{\Delta_B}{L} = \frac{19}{1536}\frac{WL^2}{EI_0} \qquad (2.5)$$

The value of W corresponding to first yield in the beam is the value at which the maximum bending moment in Figure 2.2 reaches its limiting value:

$$-M_A = M_{p0}$$

i.e.

$$18WL/64 = 0.5W_0L$$

so

$$W = 1.78W_0 \qquad (2.6)$$

It can therefore be concluded that the beam does not undergo yielding until the loads exceed their design values by 78%. The assumed moment capacity is therefore highly overconservative. Combining Equations (2.5) and (2.6), the corresponding deflection is

$$\Delta_B = \frac{19}{1536} \times 1.78W_0L^3/EI_0$$

$$= 0.0220W_0L^3/EI_0$$

and, substituting Equation (2.2),

$$\Delta_B = \frac{0.0220}{3.60}L = 0.0061L$$

2.1.2 *Elastic–plastic analysis*

Because the beam is capable of redistributing load after yielding, it has an even higher load capacity than $W = 1.78W_0$. This capacity will now be determined by carrying out an elastic–plastic load history analysis. The analysis equations will be formulated on the basis of the imposed hinge-rotation approach discussed in the introduction.

For convenience of the presentation, these equations will first be solved intuitively, then in a more general manner using the linear programming algorithm.

INTUITIVE APPROACH

For the case of the continuous beam, the sequence of hinge formation is apparent from its elastic bending moment distribution, as shown in Figure 2.2. The beam is expected to form plastic hinges at sections A, B and C, respectively, and then to fail in the form of a local simple beam mechanism in its first span. This hinge sequence is the basis of the intuitive approach which will be adopted. The analysis equations are formulated by calculating the effects of imposed hinge rotations at sections at which hinges will form. For example, referring to Figure 2.3, the effect of an imposed rotation, ϕ_A, at section A can be calculated as follows:

$$\left.\begin{aligned} M_{AC} &= \frac{2EI_0}{L}(-2\phi_A + \theta_C) \qquad M_{CA} = \frac{2EI_0}{L}(-\phi_A + 2\theta_C) \\ M_{CE} &= \frac{4EI_0}{L}\theta_C \qquad M_{EC} = \frac{2EI_0}{L}\theta_C \end{aligned}\right\} \tag{2.7}$$

From joint equilibrium at C,

$$M_{CA} + M_{CE} = 0$$

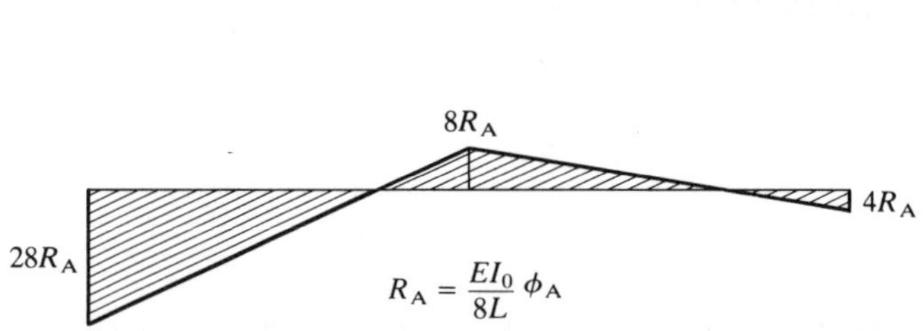

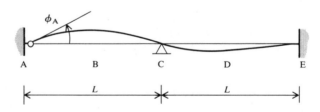

Figure 2.3 Bending moment response to an imposed hinge rotation.

i.e.

$$(2EI_0/L)(-\phi_A + 4\theta_C) = 0$$

so

$$\theta_C = \phi_A/4 \qquad (2.8)$$

Substituting Equation (2.8) into Equations (2.7), the bending moments may be expressed as

$$M_{AC} = -\frac{7}{2}\frac{EI_0}{L}\phi_A = -28R_A$$

$$M_{CA} = -M_{CE} = -\frac{EI_0}{L}\phi_A = -8R_A$$

$$M_{EC} = \frac{1}{2}\frac{EI_0}{L}\phi_A = 4R_A$$

where

$$R_A = (EI_0/8L)\phi_A$$

i.e.

$$\phi_A = (8L/EI_0)R_A \qquad (2.9)$$

The bending moments at other points along the beam can be deduced from the bending moment diagram shown in Figure 2.3. Deflections can again be deduced using the moment-area method. For example, the deflection at B is

$$\Delta_B = (\theta_A - \phi_A)\frac{L}{2} - (2M_A + M_B)\frac{L^2}{24EI_0}$$

$$= -\frac{1}{2}\phi_A L + [(2 \times 28) + 10]\frac{EI_0}{8L}\phi_A\frac{L^2}{24EI_0}$$

$$= -\frac{1}{2}\phi_A L + \frac{11}{32}\phi_A L$$

$$= -\frac{5}{32}\phi_A L$$

or, expressed in terms of the so-called hinge-rotation variable, R_A,

$$\Delta_B = -\frac{5}{32}\frac{8L}{EI_0}R_A L = -\frac{5}{4}\frac{R_A L^2}{EI_0}$$

Using the principle of superposition, the response of the beam to a combination of the loads and the hinge rotation at A can be expressed as

$$\left.\begin{aligned} &M_A = -18\frac{WL}{64} - 28R_A \qquad M_B = 17\frac{WL}{64} - 10R_A \qquad M_C = -12\frac{WL}{64} + 8R_A \\ &M_D = 7\frac{WL}{64} + 2R_A \qquad \text{and} \qquad M_E = -6\frac{WL}{64} - 4R_A \end{aligned}\right\} \quad (2.10)$$

$$24\frac{EI_0}{L^2}\Delta_B = 19\frac{WL}{64} - 30R_A \quad (2.11)$$

where

$$R_A = \frac{EI_0}{8L}\phi_A$$

A second hinge forms at section B when the bending moments at A and B are both equal to their limiting values, i.e.

$$-18\frac{WL}{64} - 28R_A = -0.5W_0L$$

$$17\frac{WL}{64} - 10R_A = 0.5W_0L$$

The solution to this set of simultaneous equations in two unknowns is

$$W = 1.85W_0 \quad (2.12)$$

$$R_A = -0.00076W_0L \quad (2.13)$$

Combining Equations (2.9) and (2.13), the hinge rotation at A is

$$\phi_A = \frac{8L}{EI_0}(-0.00076W_0L)$$

$$= -0.00608W_0L^2/EI_0$$

Substituting Equation (2.2),

$$\phi_A = -0.00608/3.6 = -0.0017\,\text{rad.}$$

The deflection at B can be determined by substituting Equations (2.12) and (2.13) into Equation (2.11). The result is

$$\Delta_B/L = 0.0239 W_0 L^2/EI_0$$

Substituting Equation (2.2),

$$\Delta_B/L = 0.0239/3.6 = 0.0066$$

Equation (2.12) reveals that the second plastic hinge forms at loads which are 85% in excess of their design values. Still further increases in load can be explored by extending Equations (2.10) and (2.11) to include the effects of an imposed rotation at the next hinge location, which is at section B. The results are found to be

$$\left.\begin{aligned} M_A &= -18\frac{WL}{64} - 28R_A - 10R_B \qquad M_B = 17\frac{WL}{64} - 10R_A - 7R_B \\ M_C &= -12\frac{WL}{64} + 8R_A - 4R_B \\ M_D &= 7\frac{WL}{64} + 2R_A - R_B \qquad M_E = -6\frac{WL}{64} - 4R_A + 2R_B \end{aligned}\right\} \quad (2.14)$$

$$24\frac{EI_0}{L^2}\Delta_B = 19\frac{WL}{64} - 30R_A + 27R_B \quad (2.15)$$

where

$$R_A = \frac{EI_0}{8L}\phi_A \qquad R_B = \frac{EI_0}{8L}\phi_B$$

As mentioned earlier, the third hinge forms at section C. The formation of this third hinge coincides with the bending moments at A, B and C all being equal to their limiting values, ie.

$$-18\frac{WL}{64} - 28R_A - 10R_B = -0.5\,W_0 L$$

$$17\frac{WL}{64} - 10R_A - 7R_B = 0.5 W_0 L$$

$$-12\frac{WL}{64} + 8R_A - 4R_B = -0.5 W_0 L$$

There are now three equations in three unknowns, and the solution is

$$W = 2.00 W_0 \qquad R_A = -0.0078 W_0 L \qquad R_B = 0.0156 W_0 L$$

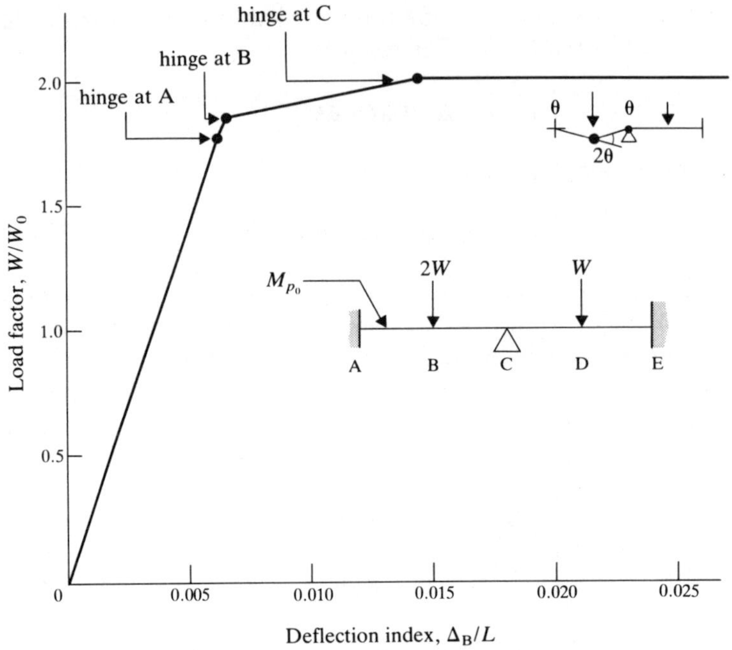

Figure 2.4 Load–deflection response.

The above values of the hinge-rotation variables refer to the state of the beam just as it forms its third hinge. Hinge rotations and deflections can be computed again as was done previously, and the results are

$$\phi_A = -0.0174\,\text{rad} \qquad \phi_B = 0.0347\,\text{rad} \qquad \Delta_B/L = 0.0145 \tag{2.16}$$

At this point, the beam forms a mechanism and collapses. The results of the analysis are summarized in the load–deflection diagram in Figure 2.4. The failure mechanism shown in the figure will be discussed subsequently.

APPLICATION OF LINEAR PROGRAMMING

In an analysis of a more complex structure, the sequence of hinge formation may not be as obvious. A more systematic approach is warranted, and one such approach is the linear programming algorithm.

Before the algorithm is applied, response is again formulated in terms of loads and hinge rotations. As indicated in Figure 2.5, rotations are imposed at all locations at which plastic hinges may occur, and the results are

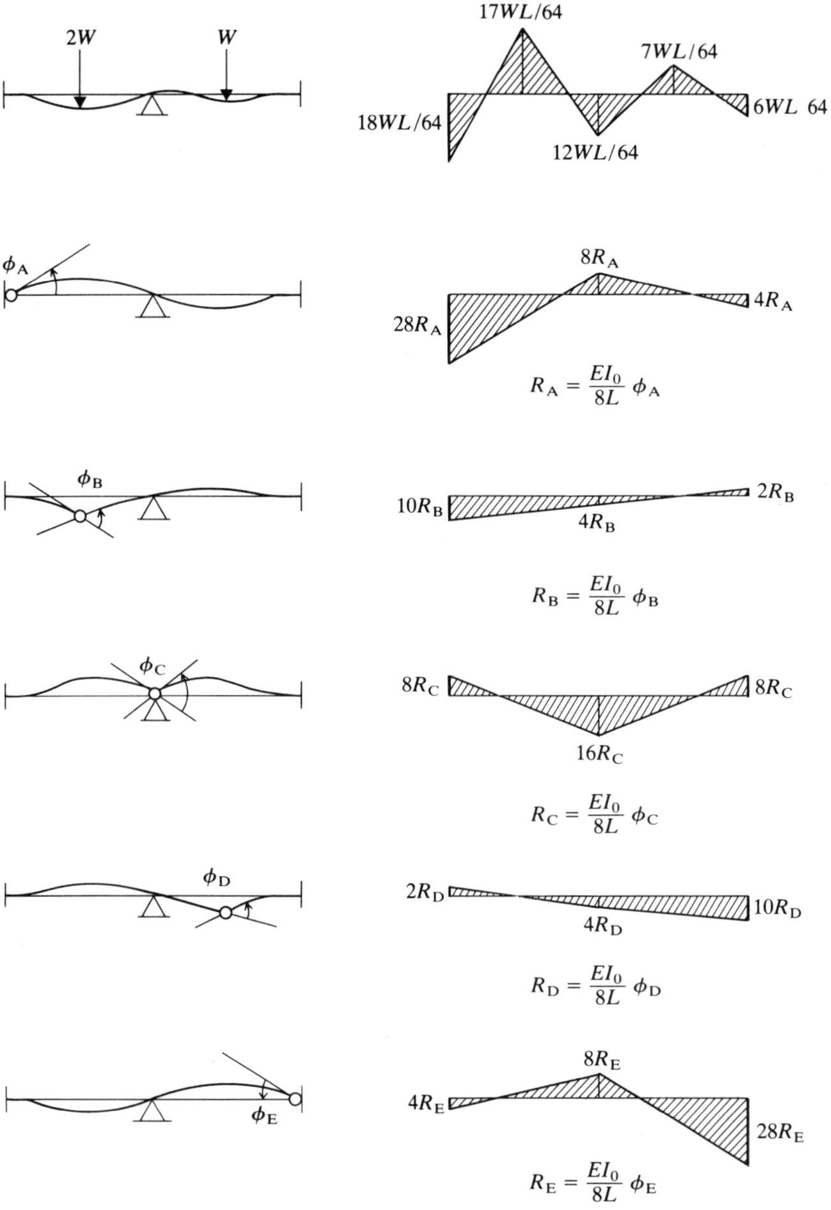

Figure 2.5 Formulation of the analysis equations.

expressed in matrix terms as follows:

$$\begin{bmatrix} M_A \\ M_B \\ M_C \\ M_D \\ M_E \end{bmatrix} = \begin{bmatrix} -18 & -28 & -10 & 8 & 2 & -4 \\ 17 & -10 & -7 & -4 & -1 & 2 \\ -12 & 8 & -4 & -16 & -4 & 8 \\ 7 & 2 & -1 & -4 & -7 & -10 \\ -6 & -4 & 2 & 8 & -10 & -28 \end{bmatrix} \begin{bmatrix} WL/64 \\ R_A \\ R_B \\ R_C \\ R_D \\ R_E \end{bmatrix} \tag{2.17}$$

$$\frac{24EI_0}{L^2} \begin{bmatrix} \Delta_B \\ \Delta_D \end{bmatrix} = \begin{bmatrix} 19 & -30 & 27 & -12 & -3 & 6 \\ 5 & 6 & -3 & -12 & 27 & -30 \end{bmatrix} \begin{bmatrix} WL/64 \\ R_A \\ R_B \\ R_C \\ R_D \\ R_E \end{bmatrix} \tag{2.18}$$

where

$$R_A = \frac{EI_0}{8L}\phi_A \qquad R_B = \frac{EI_0}{8L}\phi_B \qquad R_C = \frac{EI_0}{8L}\phi_C$$

$$R_D = \frac{EI_0}{8L}\phi_D \qquad R_E = \frac{EI_0}{8L}\phi_E$$

These equations for the bending moments and Δ_B reduce to Equations (2.14) and (2.15) for the case of $R_C = R_D = R_E = 0$. In the equations for bending moment, there is a symmetry in the coefficients of the hinge rotations. This symmetry is a direct result of the theorem of reciprocity.

The analysis problem is assumed to be one of determining the maximum loads that the beam can support without failing. Expressed numerically, the problem can be stated as follows:

$$\left.\begin{array}{l} \text{maximize } W \text{ such that:} \\ |M_A| \leqslant M_{p_0} \qquad |M_B| \leqslant M_{p_0} \qquad |M_C| \leqslant M_{p_0} \\ |M_D| \leqslant M_{p_0} \qquad |M_E| \leqslant M_{p_0} \end{array}\right\} \tag{2.19}$$

where M_A, M_B, M_C, M_D and M_E are defined in Equation (2.17), and $M_{p_0} = 0.5W_0L$.

This is an optimization problem and, because it involves linear equations only, it can be solved using the linear programming algorithm. The load parameter, W, is referred to as the objective function, the inequalities are referred to as the constraints, and the parameters W, R_A, R_B, R_C, R_D and R_E are the variables. The problem itself is referred to as a linear program.

As was evident from the discussion in the preceding section, plastic hinges occur only at sections A, B and C in the beam. The constraints and hinge rotations referring to the other two sections, D and E, can therefore be omitted in the following discussion without any loss in generality. Equation (2.19) consequently reduces to the following:

maximize W such that:

$$\begin{bmatrix} |M_A| \\ |M_B| \\ |M_C| \end{bmatrix} = \begin{bmatrix} 18 & 28 & 10 & -8 \\ 17 & -10 & -7 & -4 \\ 12 & -8 & 4 & 16 \end{bmatrix} \begin{bmatrix} WL/64 \\ R_A \\ R_B \\ R_C \end{bmatrix} \leqslant \begin{bmatrix} 0.5W_0L \\ 0.5W_0L \\ 0.5W_0L \end{bmatrix} \quad (2.20)$$

where M_A and M_C are hogging moments, and M_B is a sagging moment.

In linear programming, variables are permitted to adopt positive values only, and therefore it is necessary to ascertain the signs of the hinge rotations. As will be discussed in Section 3.5.2, the sign of a hinge rotation is a function of whether the associated bending moment is hogging or sagging. The following variable transformations are therefore warranted:

$$R_A^* = -R_A \qquad R_B^* = R_B \qquad R_C^* = -R_C$$

where R_A^*, R_B^* and R_C^* are the new hinge-rotation variables.

Equation (2.20) can therefore be rewritten as follows:

maximize W such that

$$\begin{bmatrix} 18 & -28 & 10 & 8 \\ 17 & 10 & -7 & 4 \\ 12 & 8 & 4 & -16 \end{bmatrix} \begin{bmatrix} WL/64 \\ R_A^* \\ R_B^* \\ R_C^* \end{bmatrix} \leqslant \begin{bmatrix} 0.5W_0L \\ 0.5W_0L \\ 0.5W_0L \end{bmatrix}$$

i.e.

$$\begin{bmatrix} S_A \\ S_B \\ S_C \end{bmatrix} = \begin{bmatrix} 0.5W_0L \\ 0.5W_0L \\ 0.5W_0L \end{bmatrix} - \begin{bmatrix} -18 & -28 & 10 & 8 \\ 17 & 10 & -7 & 4 \\ 12 & 8 & 4 & -16 \end{bmatrix} \begin{bmatrix} WL/64 \\ R_A^* \\ R_B^* \\ R_C^* \end{bmatrix} \geqslant \begin{bmatrix} 0 \\ 0 \\ 0 \end{bmatrix} \quad (2.21)$$

where S_A, S_B and S_C are referred to in linear programming terminology as slack variables.

The starting point for the linear programming calculations is defined by

$$W = R_A^* = R_B^* = R_C^* = 0$$

$$S_A = S_B = S_C = 0.5 W_0 L$$

The calculations are performed automatically in a series of iterations which will now be briefly described. Each iteration is characterized by what is referred to as an interchange between two of the variables.

The first iteration is characterized by an interchange between the variables, W and S_A. The value of the load parameter, W, is increased as much as possible while maintaining zero values for the hinge rotation variables, R_A^*, R_B^* and R_C^*. As indicated by Equation (2.21), the values of the slack variables must remain non-negative, and therefore the following criteria must be satisfied:

$$S_A = 0.5 W_0 L - 18 WL/64 \geqslant 0$$

$$S_B = 0.5 W_0 L - 17 WL/64 \geqslant 0$$

$$S_C = 0.5 W_0 L - 12 WL/64 \geqslant 0$$

The first criterion is the one which governs, and the maximum value of W occurs when S_A reduces to zero.

The first iteration can therefore be described as one in which S_A is reduced to zero while the load parameter, W, is increased from zero. The iteration is essentially a variable interchange in which S_A is transferred to the right-hand side of the equations, and W is transferred to the left. The calculations are the same as those of Gaussian elimination, and lead to

$$\begin{bmatrix} WL \\ S_B \\ S_C \end{bmatrix} = \begin{bmatrix} 1.78 W_0 L \\ 0.03 W_0 L \\ 0.17 W_0 L \end{bmatrix} + \begin{bmatrix} -3.56 & 99.6 & -35.6 & -28.4 \\ 0.94 & -36.4 & 16.4 & 3.6 \\ 0.67 & -26.7 & 2.7 & 21.3 \end{bmatrix} \begin{bmatrix} S_A \\ R_A^* \\ R_B^* \\ R_C^* \end{bmatrix} \qquad (2.22)$$

At the end of the first iteration $S_A = R_A^* = R_B^* = R_C^* = 0$, and the values of the variables W, S_B and S_C are defined by the vector of constants in the equation. The value of the load parameter, W, is therefore $1.78 W_0$ which, as will be recalled from Section 2.1.1, is the load required to form the first plastic hinge in the beam. The location of the hinge is section A, since it is S_A rather than S_B or S_C which reduces to zero.

Once S_A reduces to zero, it is maintained at that value while the corresponding cross section undergoes plastic deformation. In the linear program, this deformation is imposed by increasing the value of the variable R_A^*. The effect of increasing R_A^* is revealed by its coefficients in the matrix in Equation (2.22). As would be expected, it leads to an increase in W, and decreases in S_B and S_C. The second iteration involves increasing R_A^* as much as possible until one of the variables S_B or S_C reduces to zero. Since S_B reduces to zero first, this second iteration can be described as an interchange between the two variables R_A^* and S_B.

The iterations may be continued in this manner, and a summary of the calculations involved is shown in Figure 2.6. As indicated in the figure, there is a total of three iterations, each characterized by an interchange between two variables. The first is characterized by an interchange between W and S_A, the second between R_A^* and S_B, and the third between R_B^* and S_C. Each of these iterations uniquely corresponds to a step in the load history of the beam as it is gradually loaded to failure. The first corresponds to the development of a first plastic hinge at section A, the second to a subsequent plastic hinge at section B and the third to a final plastic hinge at section C. The values of the load parameter, W, as each successive hinge develops are shown in Figure 2.6 to be $1.78W_0$, $1.85W_0$ and $2.00W_0$, respectively.

Hinge rotations can be determined at any stage of loading using the

(a) Initial equations

$$\begin{bmatrix} S_A \\ S_B \\ S_C \end{bmatrix} = \begin{bmatrix} 0.5W_0L \\ 0.5W_0L \\ 0.5W_0L \end{bmatrix} + \begin{bmatrix} -18 & 28 & -10 & -8 \\ -17 & -10 & 7 & -4 \\ -12 & -8 & -4 & 16 \end{bmatrix} \begin{bmatrix} WL/64 \\ R_A^* \\ R_B^* \\ R_C^* \end{bmatrix}$$

(b) First iteration (hinge at A)

$$\begin{bmatrix} WL \\ S_B \\ S_C \end{bmatrix} = \begin{bmatrix} 1.78W_0L \\ 0.03W_0L \\ 0.17W_0L \end{bmatrix} + \begin{bmatrix} -3.6 & 99.6 & -35.6 & -28.4 \\ 0.9 & -36.4 & 16.4 & 3.6 \\ 0.7 & -26.7 & 2.7 & 21.3 \end{bmatrix} \begin{bmatrix} S_A \\ R_A^* \\ R_B^* \\ R_C^* \end{bmatrix}$$

(c) Second iteration (hinge at B)

$$\begin{bmatrix} WL \\ R_A^* \\ S_C \end{bmatrix} = \begin{bmatrix} 1.85W_0L \\ 0.0008W_0L \\ 0.15W_0L \end{bmatrix} + \begin{bmatrix} -1.0 & -2.7 & 9.4 & -18.7 \\ 0.026 & 0.027 & 0.451 & 0.097 \\ 0 & 0.7 & -9.4 & 18.7 \end{bmatrix} \begin{bmatrix} S_A \\ S_B \\ R_B^* \\ R_C^* \end{bmatrix}$$

(d) Third iteration (hinge at C)

$$\begin{bmatrix} WL \\ R_A^* \\ R_B^* \end{bmatrix} = \begin{bmatrix} 2.00W_0L \\ 0.0078W_0L \\ 0.0156W_0L \end{bmatrix} + \begin{bmatrix} -1.0 & -2.0 & -1.0 & 0 \\ 0.025 & 0.008 & -0.048 & 1.00 \\ -0.003 & 0.078 & -0.107 & 2.00 \end{bmatrix} \begin{bmatrix} S_A \\ S_B \\ S_C \\ R_C^* \end{bmatrix}$$

Figure 2.6 Load history analysis using linear programming.

following equations:

$$\phi_A = -\frac{L}{8EI_0}R_A^* \qquad \phi_B = \frac{L}{8EI}R_B^* \qquad \phi_C = -\frac{L}{8EI_0}R_C^*$$

and deflections can be calculated using Equation (2.18).

The rigid–plastic mechanism is revealed in Figure 2.6d. As indicated by the first equation, the maximum load is defined by $W = 2.00W_0$, and further increases in load are not possible without leading to negative values of the slack variables. The geometry of the mechanism is defined by the coefficients of the variable R_C^*. Once the maximum load is reached, plastic hinge rotations at A, B and C increase in the ratio of 1 : 2 : 1, respectively. This is illustrated in Figure 2.4.

2.2 Analysis–design analogy

The results of the analysis can also be used to provide design information. For example, because the load capacity is twice that for which the beam is to be designed, the moment capacity can be reduced by a factor of two. A moment capacity of $M_p = 0.5M_{p0}$ therefore corresponds to a beam which is on the verge of forming a mechanism at the design load, $W = W_0$. More generally, the analysis results in Figure 2.4 can be used to establish the design data in Figure 2.7 by inverting the ordinate function.

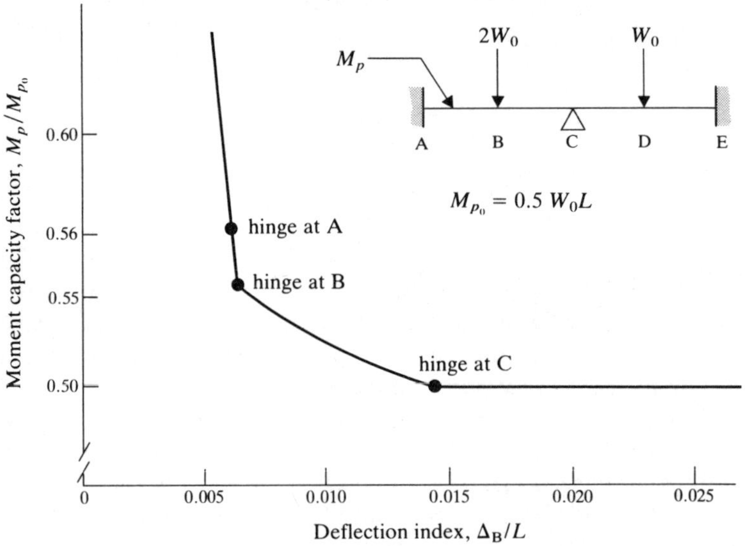

Figure 2.7 Moment capacity–deflection relationship.

For a beam of uniform cross section there is, therefore, an analogy between increasing the loads in an analysis and decreasing the moment capacity in a design. As shown in Figure 2.7, successive plastic hinges develop as the moment capacity is progressively reduced. As in the analysis, there is a total of three hinges, and they form at the following moment capacity values:

$$M_p = 0.562M_{p_0} = 0.281\,W_0L$$

$$M_p = 0.540M_{p_0} = 0.270\,W_0L$$

$$M_p = 0.500M_{p_0} = 0.250\,W_0L$$

The last value, as described earlier, refers to a beam which is on the verge of forming a mechanism at the design load. It is therefore the so-called rigid–plastic solution to the design problem. The first value, on the other hand, refers to a beam which is on the verge of forming its first plastic hinge at the design load. For a unit shape factor, this is therefore the corresponding elastic solution.

Assuming that the only criterion in the design is that the beam should not collapse before the imposed loads are reached, the design problem can be expressed as follows:

minimize M_p such that:

$$\begin{bmatrix} 18 & -28 & 10 & 8 \\ 17 & 10 & -7 & 4 \\ 12 & 8 & 4 & -16 \end{bmatrix} \begin{bmatrix} W_0L/64 \\ R_A^* \\ R_B^* \\ R_C^* \end{bmatrix} \leqslant \begin{bmatrix} M_p \\ M_p \\ M_p \end{bmatrix} \qquad (2.23)$$

where

$$R_A^* = -\frac{EI}{8L}\phi_A \qquad R_B^* = \frac{EI}{8L}\phi_B \qquad R_C^* = -\frac{EI}{8L}\phi_C$$

Design is therefore also an optimization problem which can be solved using the linear programming algorithm. In contrast to analysis, where the objective is one of maximizing loads, the objective in design is one of minimizing moment capacity.

Initially, the moment capacity is assigned an arbitrary but high value which is known to satisfy the design constraints. The value which is assumed for the purposes of this example is $M_p = W_0L$. Before applying the algorithm, the following variable transformation is warranted:

$$M_p = W_0L(1 - \mu) \qquad (2.24)$$

where μ is the new variable, and its initial value is zero. Substitution of Equation (2.24) into (2.23) leads to the following linear program.

maximize μ such that:

$$\begin{bmatrix} 1 & -28 & 10 & 8 \\ 1 & 10 & -7 & 4 \\ 1 & 8 & 4 & -16 \end{bmatrix} \begin{bmatrix} \mu \\ r_A \\ r_B \\ r_C \end{bmatrix} \leqslant \begin{bmatrix} 0.719 \\ 0.734 \\ 0.813 \end{bmatrix} \tag{2.25}$$

where

$$r_A = R_A^*/W_0L \qquad r_B = R_B^*/W_0L \qquad r_C = R_C^*/W_0L$$

Once again, three iterations are required, and these correspond to the following three sets of equations:

$$[1]\,[\mu] = [0.719]$$

$$\begin{bmatrix} 1 & -28 \\ 1 & 10 \end{bmatrix} \begin{bmatrix} \mu \\ r_A \end{bmatrix} = \begin{bmatrix} 0.719 \\ 0.734 \end{bmatrix}$$

$$\begin{bmatrix} 1 & -28 & 10 \\ 1 & 10 & -7 \\ 1 & 8 & 4 \end{bmatrix} \begin{bmatrix} \mu \\ r_A \\ r_B \end{bmatrix} = \begin{bmatrix} 0.719 \\ 0.734 \\ 0.813 \end{bmatrix}$$

These three sets of equations result in successive values of μ of 0.719, 0.730 and 0.750. The corresponding moment capacities are therefore

$$M_p = W_0L(1 - 0.719) = 0.281\,W_0L$$

$$M_p = W_0L(1 - 0.730) = 0.270\,W_0L$$

$$M_p = W_0L(1 - 0.750) = 0.250\,W_0L$$

which are the same moment capacities as were obtained earlier by applying the analysis–design analogy.

Until now, the discussion has been directed to the analysis and design of the uniform beam. For the case of the non-uniform beam in Figure 2.1b, the analysis–design analogy still applies, but is somewhat more subtle. The non-uniform beam will now be discussed, and attention will initially be confined to its elastic design.

2.3 Elastic design

As shown in Figure 2.1b, the design of the non-uniform beam is characterized by two variables, M_p and M_p', each referring to one of the two spans. For this example only, it will be assumed that the shapes of the cross sections in the two spans are the same, and are such that section properties vary in linear proportion to each other, as follows:

$$EI/M_pL = EI'/M_p'L = 7.2 \tag{2.26}$$

$$EAL/M_p = EA'L/M_p' = 5.0 \tag{2.27}$$

where M_p, I and A are the moment capacity, moment of inertia and cross-sectional area for the first span, and M_p', I' and A' are the corresponding properties in the second span.

The design constraints are assumed to be yield criteria of the form

$$|M_A| \leqslant M_p$$

where M_A is the elastic bending moment at section A and M_p refers to both the onset of yielding and the full plastification of the cross section.

As will be recalled, the elastic bending moment at A for the uniform beam is defined by

$$|M_A| = 18WL/64$$

The elastic bending moment for the non-uniform beam can be derived in a similar manner, and its value at the design loads, i.e. $W = W_0$, is

$$M_A = \frac{2(10+8n)}{1+n}\frac{W_0L}{64} \tag{2.28}$$

where

$$n = \frac{I'}{I} = \frac{M_p'}{M_p}$$

and where M_A refers to the absolute value of the bending moment.

The bending moments at the other cross sections can also be derived, and they are

$$\left.\begin{aligned} M_B &= \frac{2}{1+n}(9+8n)\frac{W_0L}{64} \qquad & M_C &= \frac{2}{1+n}(4+8n)\frac{W_0L}{64} \\ M_D &= \frac{2}{1+n}(4+3n)\frac{W_0L}{64} \qquad & M_E &= \frac{2}{1+n}(4+2n)\frac{W_0L}{64} \end{aligned}\right\} \tag{2.29}$$

where M_B, M_C, M_D and M_E refer to absolute values of bending moments.

The relationships between the bending moments and the stiffness ratio,

n, are shown graphically in Figure 2.8. As is evident from the figure, and as would be expected, increases in the value of n lead to reductions in M_A and M_B, and to increases in M_C.

The objective of the design is assumed to be one of choosing the values of M_p and M_p' corresponding to the lightest beam. Using Equation (2.27), the weight of the beam can be expressed as

$$\begin{aligned} z &= \rho L(A + A') \\ &= 5.0\rho(M_p + M_p')/E \end{aligned}$$

The objective of the design can therefore also be stated as one of minimizing the sum of the two design variables. The overall problem can therefore be expressed as follows:

$$\left.\begin{array}{ll} \multicolumn{2}{l}{\text{minimize } z = M_p + M_p' \text{ such that:}} \\ M_A = \dfrac{2}{1+n}(10+8n)\dfrac{W_0 L}{64} \leqslant M_p & M_B = \dfrac{2}{1+n}(9+8n)\dfrac{W_0 L}{64} \leqslant M_p \\ M_C = \dfrac{2}{1+n}(4+8n)\dfrac{W_0 L}{64} \leqslant M_p & M_{C'} = \dfrac{2}{1+n}(4+8n)\dfrac{W_0 L}{64} \leqslant M_p' \\ M_D = \dfrac{2}{1+n}(4+3n)\dfrac{W_0 L}{64} \leqslant M_p' & M_E = \dfrac{2}{1+n}(4+2n)\dfrac{W_0 L}{64} \leqslant M_p' \end{array}\right\} \quad (2.30)$$

where

$$n = M_p'/M_p$$

The design problem is clearly one of non-linear optimization in two variables, M_p and M_p'. Since there are only two variables, it can be solved either graphically or numerically, as will now be described.

2.3.1 *Design space*

The graphical interpretation of the problem is referred to as the design space, and is shown in Figure 2.9. As can be seen from the figure, the two axes of the space refer to the two design variables, and any point in the space represents a possible design solution. Contour A in the figure refers to section A in the beam, and represents a design yield constraint. The contour divides the design space into feasible and infeasible solutions with respect to that constraint, and passes through all of the design points for which the constraint is on the verge of being violated. The contour is derived by setting the bending moment at section A equal to its limiting

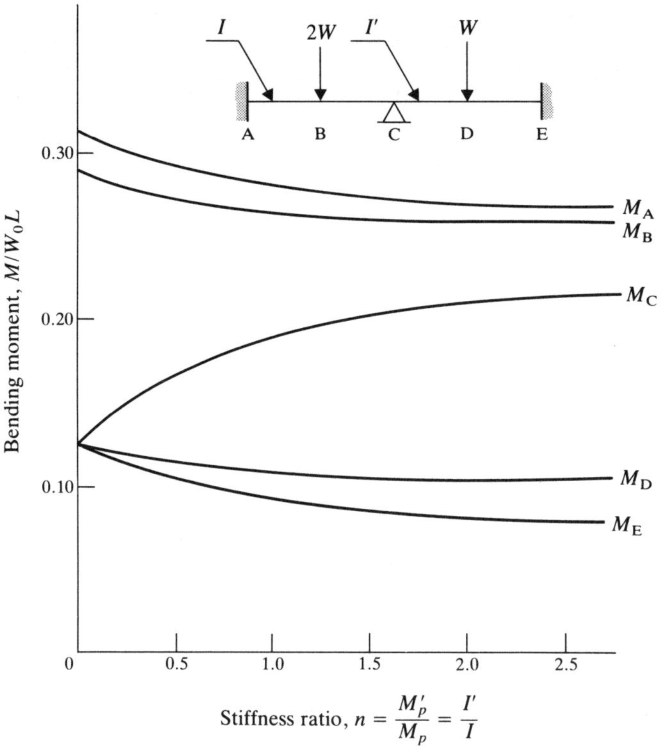

Figure 2.8 Relationship between bending moments and stiffness ratio.

value, i.e.

$$\frac{2}{1+n}(10+8n)\frac{W_0L}{64}=M_p$$

where

$$n=M_p'/M_p$$

Other contours can be derived in a similar manner. Contours B and C, for example, are derived by setting the bending moments at the corresponding sections equal to their limiting values. They are not shown in Figure 2.9 because they lie entirely in the infeasible inelastic domain, and are therefore not relevant. As indicated in Figure 2.8, the bending moments at sections B and C are always less than the bending moment at section A. Section A is therefore the only location in the first span at which first yield can occur.

As is also indicated in Figure 2.8, the only location in the second span at which first yield can occur is at the mid-support. The corresponding

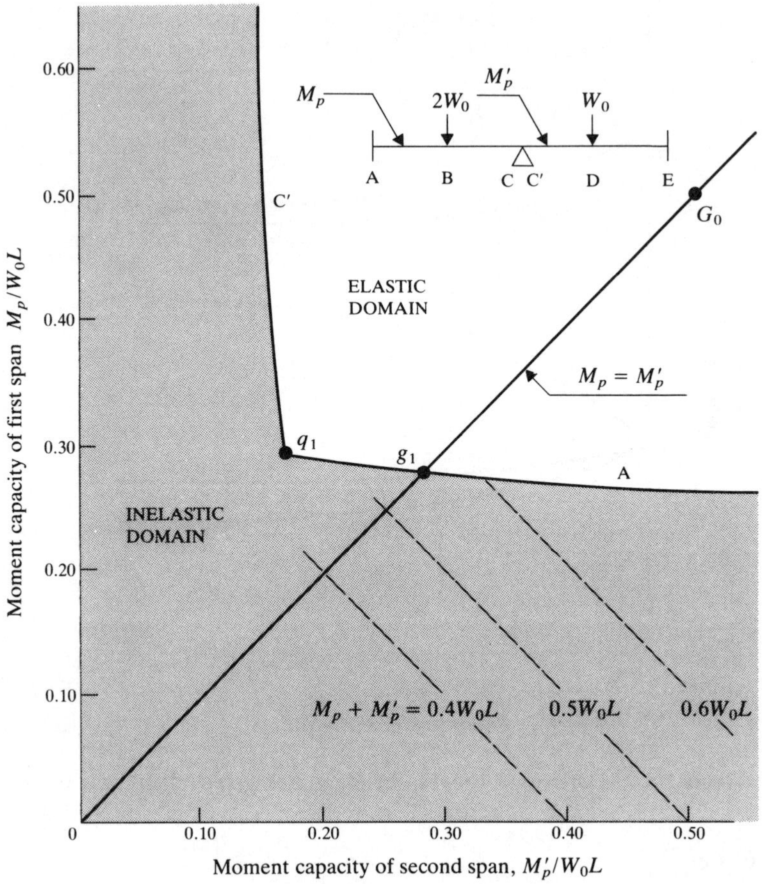

Figure 2.9 The elastic design space.

contour is designated C′ to refer to the section immediately to the right of the support, and is derived by setting the bending moment at C equal to its limiting value, i.e.

$$\frac{2}{1+n}(4+8n)\frac{W_0L}{64} = M_p'$$

As shown in Figure 2.9, the two contours together divide the design space into two domains; one which defines design solutions which remain elastic under the imposed loads, and one which defines solutions which experience yielding.

The system of parallel lines in Figure 2.9 defines the design objective, which, as will be recalled, is one of minimizing the sum of the moment capacities. As can be seen from the figure, the minimum value which can be achieved without penetrating the inelastic domain corresponds to the point q_1. The optimum solution to the design problem is therefore

$$M_p = 0.289 W_0 L \qquad M'_p = 0.171 W_0 L \tag{2.31}$$

Since the point q_1 coincides with the intersection of two yield constraints, the beam yields at both sections A and C′ simultaneously at the instant the imposed loads are reached. The solution is therefore a fully stressed elastic design.

2.3.2 Design calculations

A graphical approach for solving the optimization problem represented by Equation (2.30) is clearly limited in application. There are generally more than two variables involved, and a numerical technique is warranted. The proposed plastic design method is such a technique, and it reduces to a method of elastic design simply by omitting to specify the plastic hinge-rotation variables.

This approach, which for elastic design is the same as that adopted by Reinschmidt *et al.* (1966), is one in which the non-linear equations involved are solved in an iterative manner. An initial estimate of the design solution is chosen, and is progressively improved in a series of design cycles until convergence is obtained. Each cycle involves two operations: formulation and optimization. Formulation involves reducing the non-linear optimization problem to an approximate linear one on the basis of the current estimate of the design solution. Optimization involves solving the linear approximation using linear programming. Both operations will now be described for the continuous beam example.

FORMULATION

The linear approximation to the optimization problem in Equation (2.30) is its first-order Taylor expansion with respect to the current trial design solution, i.e.

minimise $z = M_p + M'_p$ subject to constraints of the form:

$$M_{A_0} + \frac{dM_A}{dn}\bar{n} \leqslant M_p$$

where

$$n = M'_p/M_p \qquad n_0 = M'_{p0}/M_{p0} \qquad \bar{n} = n - n_0 \tag{2.32}$$

and where M_{p_0} and M'_{p_0} define the current design solution.

The formulation operation is therefore essentially one of deriving linear approximations to the equations for the expressions for the bending moments. The bending moment at section A, for example, is

$$M_A = \frac{2}{1+n}(10+8n)\frac{W_0L}{64} \tag{2.33}$$

Its linear approximation is

$$M_A = M_{A_0} + \frac{dM_A}{dn}\bar{n} \tag{2.34}$$

where

$$M_{A_0} = \frac{2}{1+n_0}(10+8n_0)\frac{W_0L}{64}$$

$$\frac{dM_A}{dn} = -\frac{4}{(1+n_0)^2}\frac{W_0L}{64}$$

In the first design cycle, the current design solution is an arbitrary one, and is assumed to be the uniform beam solution introduced in Section 2.1, i.e.

$$M_{p_0} = M'_{p_0} = 0.5W_0L$$

$$n_0 = M'_{p_0}/M_{p_0} = 1.0$$

Equation (2.34) therefore reduces to

$$M_A = (18-\bar{n})\frac{W_0L}{64} \tag{2.35}$$

This equation is the linear approximation of Equation (2.33) with respect to the initial design, and both equations are illustrated in Figure 2.10 where they are designated a and b. The approximation is clearly a significant improvement on the assumption, as represented by c in the figure, that the bending moment remains constant with design changes. This last assumption is typically the basis of manual design methods.

The linear approximation to the yield constraint at section A is therefore

$$(18-\bar{n})W_0L/64 \leqslant M_p \tag{2.36}$$

Since $n = M'_p/M_p$, the parameter $\bar{n}$ can be expressed explicitly in terms of the design variables:

$$\bar{n} = \frac{\partial n}{\partial M_p}\overline{M}_p + \frac{\partial n}{\partial M'_p}\overline{M}'_p$$

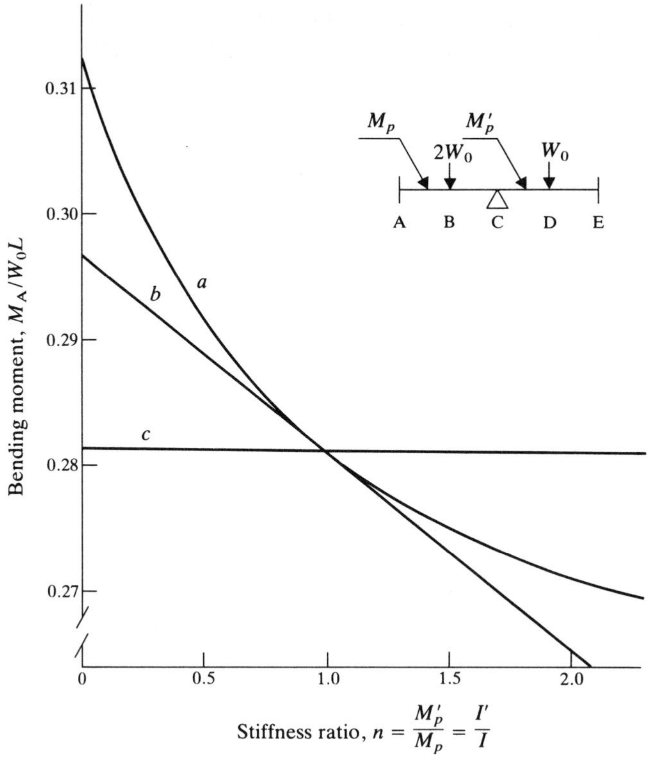

Figure 2.10 Linearization of a response function.

where

$$\frac{\partial n}{\partial M_p} = -M'_{p0}/(M_{p0})^2 = -\frac{2}{W_0 L}$$

$$\frac{\partial n}{\partial M'_p} = 1/M_{p0} = \frac{2}{W_0 L}$$

$$\bar{M}_p = M_p - M_{p0} = M_p - 0.5W_0 L$$

$$\bar{M}'_p = M'_p - M'_{p0} = M'_p - 0.5W_0 L$$

i.e.

$$\bar{n} = \frac{2}{W_0 L}(\bar{M}'_p - \bar{M}_p)$$

$$= \frac{2}{W_0 L}(M'_p - M_p) \tag{2.37}$$

Constraint (2.36) can therefore be rewritten as

$$18\frac{W_0L}{64}-\frac{2}{64}(M_p'-M_p)\leqslant M_p$$

i.e.

$$0.969M_p+0.031M_p'\geqslant 18W_0L/64$$

The other constraints in Equation (2.30) can be linearized in a similar manner, and the optimization problem reduces to:

minimize $z=M_p+M_p'$ such that:

$$\begin{bmatrix}0.969 & 0.031\\ 0.984 & 0.016\\ 1.063 & -0.063\\ 0.063 & 0.938\\ -0.016 & 1.016\\ -0.031 & 1.031\end{bmatrix}\begin{bmatrix}M_p\\ M_p'\end{bmatrix}\geqslant\begin{bmatrix}18\\ 17\\ 12\\ 12\\ 7\\ 6\end{bmatrix}[W_0L/64] \tag{2.38}$$

This is a standard linear optimization problem, which can be solved using the linear programming algorithm.

OPTIMIZATION

As was the case in Section 2.2, the linear programming calculations are begun by carrying out a variable transformation. A convenient transformation is

$$M_p=W_0L(1-\mu)\qquad M_p'=W_0L(1-\mu-\nu) \tag{2.39}$$

where the starting point for the calculations is defined by zero values for the two new variables, μ and ν. The transformation above is chosen so that proportional changes in the moment capacities can be investigated separately by varying μ before changing the value of ν.

Substituting Equations (2.39) into Equations (2.38) and introducing slack variables, the linear program can be rewritten as:

minimize $z=2-2\mu-\nu$ such that:

$$\begin{bmatrix}1.000 & 0.031\\ 1.000 & 0.016\\ 1.000 & -0.063\\ 1.000 & 0.938\\ 1.000 & 1.016\\ 1.000 & 1.031\end{bmatrix}\begin{bmatrix}\mu\\ \nu\end{bmatrix}+\begin{bmatrix}S_A\\ S_B\\ S_C\\ S_{C'}\\ S_D\\ S_E\end{bmatrix}=\begin{bmatrix}0.719\\ 0.734\\ 0.813\\ 0.813\\ 0.891\\ 0.906\end{bmatrix} \tag{2.40}$$

As indicated in the above equation, there are a total of six slack variables – one for each of the yield constraints. The optimization problem is reduced to one of minimizing the value of z while maintaining non-negative values for all of the slack variables.

The solution to the problem can be obtained in a manner similar to that described in Section 2.2 for elastic–plastic analysis. In the case of design, each step in the linear programming calculations corresponds to a specific movement in the design space in Figure 2.9. As indicated in the figure, only two movements, G_0g_1 and g_1q_1, are required, and both will now be briefly outlined.

The starting point, G_0, lies along the line $M_p = M_p'$, and is defined by zero values for the two variables μ and ν. In the first step, the value of μ is increased until one of the slack variables, S_A, reduces to zero. The step leads to a new point g_1, which is defined by zero values for the two variables ν and S_A. Referring to Equation (2.40), the corresponding solution is

$$[1.000]\,[\mu] = [0.719]$$

i.e.

$$\mu = 0.719 \qquad \nu = 0.000$$

i.e.

$$M_p = W_0L(1 - 0.719) = 0.281\,W_0L$$

$$M_p' = W_0L(1 - 0.719 - 0.000) = 0.281\,W_0L$$

In the second step, the value of ν is increased until a second slack variable, $S_{C'}$, reduces to zero. The step leads to the point q_1, which is defined by zero values for the two variables S_A and $S_{C'}$. Referring to Equation (2.40), the corresponding solution is obtained as follows:

$$\begin{bmatrix} 1.000 & 0.031 \\ 1.000 & 0.938 \end{bmatrix} \begin{bmatrix} \mu \\ \nu \end{bmatrix} = \begin{bmatrix} 0.719 \\ 0.813 \end{bmatrix}$$

i.e.

$$\mu = 0.716 \qquad \nu = 0.103$$

i.e.

$$M_p = W_0L(1 - 0.716) = 0.284\,W_0L$$

$$M_p' = W_0L(1 - 0.716 - 0.103) = 0.181\,W_0L$$

Because of linearization errors, this last step results only in an approximate estimate of the solution, corresponding to the point q_1. If this estimate is used as the next trial design in a new cycle of formulation and optimization, the solution obtained would be

$$M_p = 0.289 W_0 L \qquad M_p' = 0.171 W_0 L$$

which is the required solution.

2.4 Plastic design

The non-uniform example in Fig. 2.1b can also be used to discuss plastic design. Once again, the design problem is defined in terms of its variables, objective and constraints. The main variables are again the moment capacities M_p and M_p'. The shape of the cross sections are assumed to satisfy the same relationships, (2.26) and (2.27), and the objective is again assumed to be one of minimizing the sum of the moment capacities.

The major difference, as was the case in plastic analysis, is the need for the introduction of hinge-rotation variables.

The constraints in a plastic design may comprise any combination of yield, ductility or serviceability criteria. These criteria may be expressed in the forms

$$|M| \leqslant M_p \qquad |\phi| \leqslant \phi_{\max} \qquad |\Delta| \leqslant \Delta_{\max} \tag{2.41}$$

where M and ϕ are a bending moment and hinge rotation under ultimate load, and Δ is a deflection under serviceability load.

In the elastic design of the beam, response was expressed in terms of the load parameter W and the stiffness ratio n. In plastic design, the same expressions may be extended to include the effects of plastic hinge rotations at the six cross sections, A, B, C, C′, D and E, in the beam. As will be seen, plastic hinge rotations do not occur at sections C, D and E, and therefore the associated variables can be omitted without loss in generality. The expressions for the response are

$$\begin{bmatrix} M_A \\ M_B \\ M_C \\ M_{C'} \\ M_D \\ M_E \end{bmatrix} = \frac{2}{1+n} \begin{bmatrix} 10+8n & -(12+16n) & 6+4n & 8 \\ 9+8n & 6+4n & -(3+4n) & 4 \\ 4+8n & 8n & 4n & -16 \\ 4+8n & 8n & 4n & -16 \\ 4+3n & -2n & -n & 4 \\ 4+2n & -4n & -2n & 8 \end{bmatrix} \begin{bmatrix} W_0 L/64 \\ R_A \\ R_B \\ R_C \end{bmatrix} \tag{2.42}$$

$$\frac{24EI}{L^2}\begin{bmatrix}\Delta_B\\ \Delta_D\end{bmatrix} = \frac{2}{1+n}\begin{bmatrix}11+8n & 18+12n & 15+12n & 12\\ 1+4/n & -6 & -3 & 12/n\end{bmatrix}\begin{bmatrix}W_0L/64\\ R_A\\ R_B\\ R_{C'}\end{bmatrix} \qquad (2.43)$$

where

$$R_A = -\frac{EI}{8L}\phi_A \qquad R_B = \frac{EI}{8L}\phi_B$$
$$R_{C'} = -\frac{EI'}{8L}\phi_{C'} \qquad n = M_p'/M_p \qquad (2.44)$$

and where the bending moments are defined in terms of their absolute values. The above equations reduce to Equations (2.17) and (2.18) for the case of a unit stiffness ratio n.

In spite of the fact that the design problem now involves more than two variables, it can still be represented graphically, as will now be seen. Application of the numerical approach will be described subsequently.

2.4.1 Design space

Once the loads are known, the response of the beam is defined uniquely in terms of the two independent design variables M_p and M_p'. The hinge-rotation variables are by no means independent, as they are themselves functions of M_p and M_p'. Contours can therefore be drawn to represent yield, serviceability or ductility criteria, all on the one design space.

YIELD CRITERIA

As will be recalled, an elastic yield contour defines design solutions for which first yield just occurs at the instant the design loads are reached. Referring to Equation (2.42), the yield contour A in Figure 2.9, for example, is therefore defined by

$$\frac{2}{1+n}(10+8n)\frac{W_0L}{64} = M_p$$

A plastic yield contour is one which refers to the beam undergoing a more advanced stage of yielding at the instant the applied loads are reached. The contour AB in Figure 2.11, for example, corresponds to design solutions for which a second hinge forms at section B after hinge rotation has already occurred at section A. The relevant equations can also be derived from

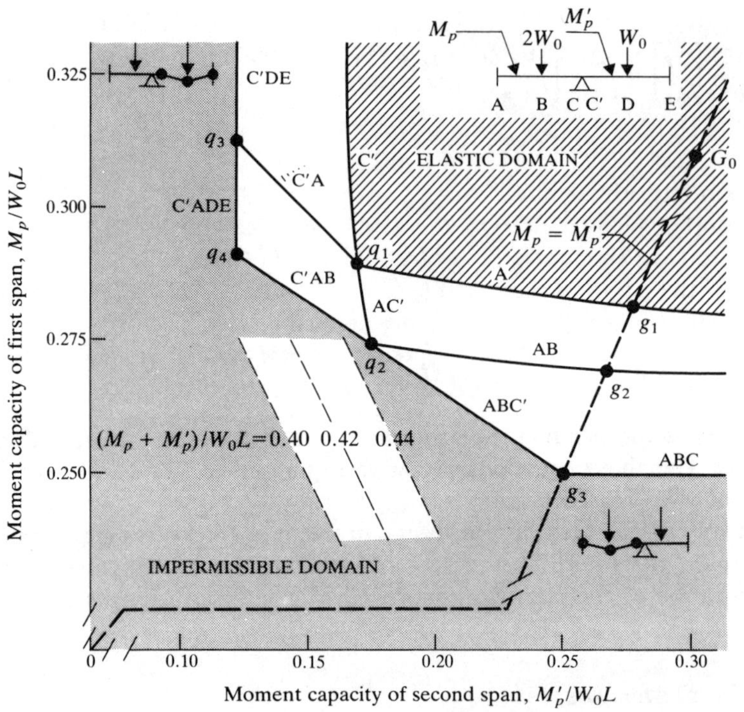

Figure 2.11 The design space.

Equation (2.42), and are

$$\frac{2}{1+n}\begin{bmatrix} 10+8n & -(12+16n) \\ 9+8n & 6+4n \end{bmatrix}\begin{bmatrix} W_0L/64 \\ R_A \end{bmatrix} = \begin{bmatrix} M_p \\ M_p \end{bmatrix}$$

Elimination of R_A leads to

$$\frac{21+20n}{18+20n}\,\frac{W_0L}{4} = M_p$$

which is the equation corresponding to the contour AB in the figure.

Contour ABC′ in Figure 2.11 can also be derived, and defines design solutions for which a third hinge forms at section C′ after hinge rotation has occurred at sections A and B. From Equation (2.42), the relevant equations are

$$\frac{2}{1+n}\begin{bmatrix} 10+8n & -(12+16n) & 6+4n \\ 9+8n & 6+4n & -(3+4n) \\ 4+8n & 8n & 4n \end{bmatrix}\begin{bmatrix} W_0L/64 \\ R_A \\ R_B \end{bmatrix} = \begin{bmatrix} M_p \\ M_p \\ M'_p \end{bmatrix}$$

Elimination of R_A and R_B leads to

$$3M_p + M_p' = W_0 L$$

Because contour ABC′ describes a rigid–plastic mechanism condition, this last equation is characteristically independent of the parameter n, and is linear in terms of the two design variables.

The other yield contours in the design space can be derived in a similar manner. Their significance is apparent in their relationship to lines passing through the origin of the space. The line $M_p = M_p'$, for example, refers to the specific case of the uniform beam discussed earlier. As shown in Figure 2.11, the line intersects the contours A, AB and ABC, and the intersection points are g_1, g_2 and g_3, respectively. These three points correspond to the successive stages in hinge development, both in the proportional design of the beam in Figure 2.7 and in its analogous load history analysis in Figure 2.4.

The last point, g_3, refers to a uniform beam which is on the verge of forming a mechanism at the design loads. For the more general case of a non-uniform beam, points referring to a mechanism at the design loads all lie on the so-called rigid–plastic contours; C′DE, C′ADE, C′AB, ABC′ and ABC. These contours are linear, and in a rigid–plastic design they define the boundary separating permissible and impermissible solutions. Referring to the lines of constant weight in Figure 2.11, they also define the optimum design solution at the point q_4. As will be discussed in Section 2.4.2 with reference to Figure 2.13, this solution is characterized by plastic hinges at five possible locations along the beam, namely, A, B, C′, D and E. The solution can be easily verified by applying rigid–plastic theory, and its relative economy is compared to that of other permissible solutions in the design space in Table 2.1.

Table 2.1 Four alternative design solutions.

	q_1	q_2	q_3	q_4
M_p/W_0L	0.289	0.274	0.313	0.292
M_p'/W_0L	0.171	0.177	0.125	0.125
$(M_p + M_p')/W_0L$	0.460	0.451	0.438	0.417

SERVICEABILITY AND DUCTILITY CRITERIA

If serviceability or ductility criteria are to be included in the design, the solution corresponding to the point q_4 may not be valid.

For the purpose of defining serviceability criteria, it is necessary to introduce a load factor λ_0 such that

$$W_0 = \lambda_0 w_0 \qquad (2.45)$$

where W_0 and w_0 represent ultimate and working load levels, respectively. For the purposes of this example, it is arbitrarily assumed that the load factor λ_0 is equal to 1.5, and the serviceability criterion is the deflection constraint:

$$|\Delta_B| \leqslant L/240 \tag{2.46}$$

As can be shown for this example, working load response is elastic. Using Equations (2.43) and (2.45), the deflection Δ_B at working load is therefore

$$|\Delta_B| = \frac{2(11+8n)}{1+n}\frac{W_0L}{64}\frac{L^2}{24EI}\frac{1}{1.5} = \frac{2(11+8n)}{1+n}\frac{W_0L^3}{2304EI}$$

The contour corresponding to the serviceability constraint can be derived by setting the deflection Δ_B equal to its limiting value, $L/240$. Substituting Equation (2.26), the equation for the contour is

$$\frac{2(11+8n)}{1+n}\frac{W_0L^2}{2304 \times 7.2M_p} = \frac{L}{240}$$

i.e.

$$\frac{2(11+8n)}{1+n}\frac{W_0L}{69.12} = M_p$$

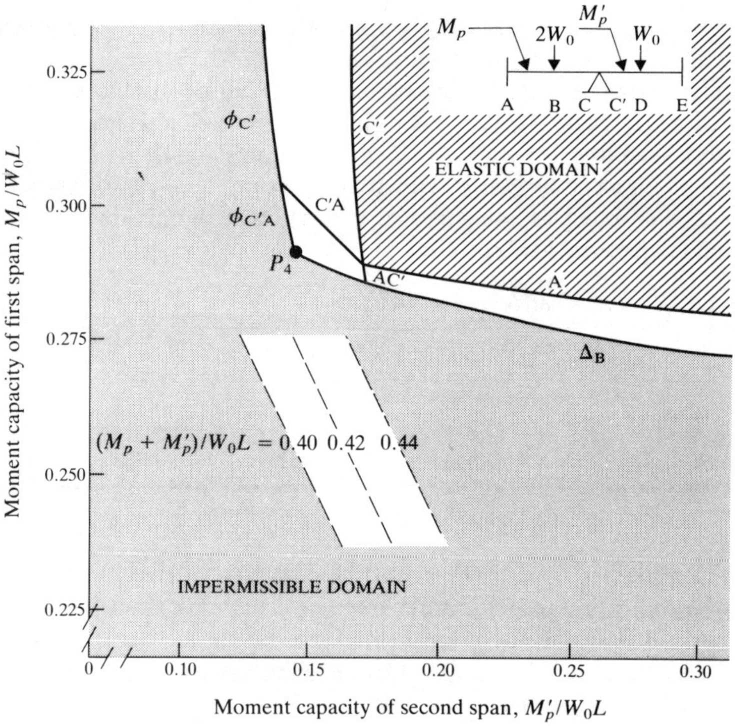

Figure 2.12 The design space with deflections and ductility constraints.

The above equation is represented in Figure 2.12 as contour Δ_B. Points on the contour represent design solutions on the verge of violating the serviceability criterion. The contour has no relationship with yielding in the beam, and is therefore continuous at intersection points with the various yield contours.

The ductility criteria are arbitrarily assumed to be as follows:

$$|\phi_A| \leqslant 0.01 \text{ rad} \qquad |\phi_B| \leqslant 0.01 \text{ rad} \qquad |\phi_{C'}| \leqslant 0.01 \text{ rad} \tag{2.47}$$

The first step in the derivation of the associated contours is one of deriving expressions for the hinge rotations in terms of the design and hinge-rotation variables. Using Equations (2.26) and (2.44), these expressions are

$$\left.\begin{aligned} \phi_A &= -\frac{8L}{EI} R_A = -\frac{8}{7.2M_p} R_A = -1.11\frac{R_A}{M_p} \\ \phi_B &= \frac{8L}{EI} R_B = 1.11\frac{R_B}{M_p} \\ \phi_{C'} &= -\frac{8L}{EI'} R_{C'} = -1.11\frac{R_{C'}}{M_p'} \end{aligned}\right\} \tag{2.48}$$

The next step is to set the hinge rotations equal to their limiting values. For the ductility constraint for section C′, for example, the equation is

$$1.11R_{C'}/M_p' = 0.01$$

i.e.

$$R_{C'} = 0.009M_p' \tag{2.49}$$

As can be deduced from Figure 2.11, the state of yielding at which this limiting value is reached is not unique, and depends on the design solution. There are, in fact, two possibilities. If the design solution lies between the contours C′, C′A and C′DE, the state of yielding is one in which there is a plastic hinge only at section C′. The associated contour equation can be derived as follows:

$$\begin{bmatrix} M_{C'} \\ R_{C'} \end{bmatrix} = \begin{bmatrix} M_p' \\ 0.009M_p' \end{bmatrix}$$

i.e.

$$\frac{2}{1+n}\begin{bmatrix} 4+8n & -16 \\ 0 & (n+1)/2 \end{bmatrix}\begin{bmatrix} W_0L/64 \\ R_{C'} \end{bmatrix} = \begin{bmatrix} M_p' \\ 0.009M_p' \end{bmatrix}$$

Elimination of $R_{C'}$ leads to

$$M_p' = \frac{W_0L}{4}\frac{n+0.500}{n+1.288}$$

which is represented by the contour $\phi_{C'}$ in Figure 2.12.

Alternatively, if the design solution lies between C′A, AC′, C′AB and C′ADE, two plastic hinges form before the limiting rotation is reached. The associated contour equation can be derived as follows:

$$\begin{bmatrix} M_A \\ M_{C'} \\ R_{C'} \end{bmatrix} = \begin{bmatrix} M_p \\ M_p' \\ 0.009M_p' \end{bmatrix}$$

i.e.

$$\frac{2}{1+n}\begin{bmatrix} 10+8n & -(12+16n) & 8 \\ 4+8n & 8n & -16 \\ 0 & 0 & (n+1)/2 \end{bmatrix}\begin{bmatrix} W_0L/64 \\ R_A \\ R_{C'} \end{bmatrix} = \begin{bmatrix} M_p \\ M_p' \\ 0.009M_p' \end{bmatrix}$$

Elimination of R_A and $R_{C'}$ yields

$$M_p' = \frac{3W_0L}{8} = \frac{n+0.250}{n+1.466}$$

which is represented by the contour $\phi_{C'A}$ in Figure 2.12.

The ductility criterion at section C′ is therefore represented in the design space by two contours, $\phi_{C'}$ and $\phi_{C'A}$. Contours for the ductility criteria at the other cross sections in the beam are not shown in Figure 2.12 since they have no effect on the design solution. They are derived in a similar manner, and lie entirely in the impermissible domain.

As can be deduced from the lines of equal weight in the figure, the solution to the design problem is the one corresponding to the point p_4 at the intersection of contours Δ_B and $\phi_{C'A}$. The corresponding values of the moment capacities are

$$M_p = 0.290W_0L \qquad M_p' = 0.142W_0L$$

2.4.2 *Design calculations*

As discussed earlier, a graphical approach is impractical for all but the simplest structures, and design solutions may be determined numerically using the proposed plastic design approach. As in the application of the approach to elastic design problems, an initial estimate of the design solution is chosen, and this is progressively improved in a series of design cycles until convergence is obtained. Again, each cycle involves the two operations of formulation and optimization. The procedure will now be described for the continuous beam example and, for convenience of presentation, attention will be directed to a design which is subject only to yield constraints.

FORMULATION

In the case of the elastic design of the beam, formulation involved linearizing the response equation with respect to the stiffness ratio *n*. In plastic design, on the other hand, the response equations are linearized not only with respect to the ratio *n*, but also with respect to the hinge rotation variables. The linearized equation for the bending moment response at section A, for example, may be written as

$$M_A = M_{A_0} + \frac{\partial M_A}{\partial n}\bar{n} + \frac{\partial M_A}{\partial R_A}\bar{R}_A + \frac{\partial M_A}{\partial R_B}\bar{R}_B + \frac{\partial M_A}{\partial R_{C'}}\bar{R}_{C'} \tag{2.50}$$

where

$$n = M_p'/M_p \qquad \bar{n} = n - n_0$$

$$\bar{R}_A = R_A - R_{A_0} \qquad \bar{R}_B = R_B - R_{B_0} \qquad \bar{R}_{C'} = R_{C'} - R_{C_0'}$$

The initial values of the hinge-rotation variables are generally assumed to be zero, and the initial estimate of the design solution may once again be defined by

$$M_{p_0} = M_{p_0}' = 0.5\,W_0 L$$

$$(2.51)$$

$$n_0 = M_{p_0}'/M_{p_0} = 1.0$$

The values of M_{A_0} and the partial derivatives of M_A in Equation (2.50) are derived from Equation (2.42), and are as follows:

$$M_{A_0} = \frac{2}{1+n_0}(10 + 8n_0)\frac{W_0 L}{64} = 18\,\frac{W_0 L}{64}$$

$$\frac{\partial M_A}{\partial n} = -\frac{4}{(1+n_0)^2}\left(\frac{W_0 L}{64} + 2R_{A_0} + R_{B_0} + 4R_{C_0'}\right) = -\frac{W_0 L}{64}$$

$$\frac{\partial M_A}{\partial R_A} = -\frac{2}{1+n_0}(12 + 16n_0) = -28$$

$$\frac{\partial M_A}{\partial R_B} = \frac{2}{1+n_0}(6 + 4n_0) = 10$$

$$\frac{\partial M_A}{\partial R_{C'}} = \frac{2}{1+n_0}(8) = 8$$

The linear approximation (2.50) to the bending moment equation therefore becomes

$$M_A = (18 - \bar{n})\frac{W_0 L}{64} - 28R_A + 10R_B + 8R_{C'}$$

Substituting Equation (2.37) for the parameter $\bar{n}$, the associated yield constraint can be written as

$$M_A = 18\frac{W_0 L}{64} - \frac{2}{64}(M_p' - M_p) - 28R_A + 10R_B + 8R_{C'} \leqslant M_p$$

i.e.

$$0.969M_p + 0.031M_p' + 28R_A - 10R_B - 8R_{C'} \geqslant 18W_0L/64$$

The other yield constraints can be linearized in a similar manner, and the design problem reduces to:

minimize $z = M_p + M_p'$ such that:

$$\begin{bmatrix} 0.969 & 0.031 & 28 & -10 & -8 \\ 0.984 & 0.016 & -10 & 7 & -4 \\ 1.063 & -0.063 & -8 & -4 & 16 \\ 0.063 & 0.938 & -8 & -4 & 16 \\ -0.016 & 1.016 & 2 & 1 & -4 \\ -0.031 & 1.031 & 4 & 2 & -8 \end{bmatrix} \begin{bmatrix} M_p \\ M_p' \\ R_A \\ R_B \\ R_{C'} \end{bmatrix} \geqslant \begin{bmatrix} 18 \\ 17 \\ 12 \\ 12 \\ 7 \\ 6 \end{bmatrix} [W_0L/64] \quad (2.52)$$

Once again, this is a linear optimization problem, which can be solved using the linear programming algorithm.

OPTIMIZATION

In the linear programming calculations, the same variable transformations are adopted as for elastic design, i.e.

$$M_p = 2M_{p0}(1 - \mu)$$
$$M_p' = 2M_{p0}'(1 - \mu - \nu) \quad (2.53)$$

where

$$M_{p0} = M_{p0}' = 0.5W_0L$$

Substituting the above equations into Equations (2.52), the linear program becomes:

minimize $z = 2 - 2\mu - \nu$ such that:

$$\begin{bmatrix} 1.000 & 0.031 & -28 & 10 & 8 \\ 1.000 & 0.016 & 10 & -7 & 4 \\ 1.000 & -0.063 & 8 & 4 & -16 \\ 1.000 & 0.938 & 8 & 4 & -16 \\ 1.000 & 1.016 & -2 & -1 & 4 \\ 1.000 & 1.031 & -4 & -2 & 8 \end{bmatrix} \begin{bmatrix} \mu \\ \nu \\ r_A \\ r_B \\ r_{C'} \end{bmatrix} \leqslant \begin{bmatrix} 0.719 \\ 0.734 \\ 0.813 \\ 0.813 \\ 0.891 \\ 0.906 \end{bmatrix} \tag{2.54}$$

where

$$r_A = R_A/W_0L \qquad r_B = R_B/W_0L \qquad r_{C'} = R_{C'}/W_0L$$

The linear program is thus expressed in terms of the five variables μ, ν, r_A, r_B and $r_{C'}$ all of which are initially equal to zero. Slack variables, S_A, S_B, S_C, $S_{C'}$, S_D and S_E, are introduced for the six constraints in the same manner as in the case of elastic design.

The determination of the solution to the program is divided into two phases, both of which will now be described. In the first phase, the variable ν is maintained at a value of zero. The equations of the linear program effectively reduce to Equations (2.25) for proportional design, and the calculations are analogous to those of the load history analysis for the current design solution. A total of three iterations is involved, and these iterations correspond to the three design movements G_0g_1, g_1g_2 and g_2g_3 in the design space in Figure 2.11. In terms of the linear programming algorithm, they represent interchanges between variables μ and S_A, r_A and S_B, and r_B and $S_{C'}$, respectively. At the end of the first phase, the values of S_A, S_B, $S_{C'}$, ν and $r_{C'}$ are equal to zero, and the values of μ, r_A and r_B are defined by

$$\begin{bmatrix} 1.000 & -28 & 10 \\ 1.000 & 10 & -7 \\ 1.000 & 8 & 4 \end{bmatrix} \begin{bmatrix} \mu \\ r_A \\ r_B \end{bmatrix} = \begin{bmatrix} 0.719 \\ 0.734 \\ 0.813 \end{bmatrix}$$

In the second phase of the linear programming calculations, improvements in the value of the objective function are achieved by allowing the value of ν to vary freely. Two iterations are involved, and the first corresponds to the movement g_3q_2 in Figure 2.11. Since the movement is a departure from the line $M_p = M_p'$, the variable ν is increased from its zero value. Since it ends at the yield contour AB, the variable r_B is reduced to zero. The movement therefore corresponds to an interchange between ν and r_B in the linear programming equations, and the new point q_2 is defined by the equations

$$\begin{bmatrix} 1.000 & 0.031 & -28 \\ 1.000 & 0.016 & 10 \\ 1.000 & 0.938 & 8 \end{bmatrix} \begin{bmatrix} \mu \\ \nu \\ r_A \end{bmatrix} = \begin{bmatrix} 0.719 \\ 0.734 \\ 0.813 \end{bmatrix}$$

The second iteration corresponds to the movement q_2q_4 in Figure 2.11. Since the movement is a departure from the contour AC′, the variable $r_{C'}$ is increased from its zero value, and since it ends at the yield contour C′AD, the slack variable S_D is reduced to zero. The movement therefore corresponds to an interchange between $r_{C'}$ and S_D, and the new point q_4 is defined by the equations

$$\begin{bmatrix} 1.000 & 0.031 & -28 & 8 \\ 1.000 & 0.016 & 10 & 4 \\ 1.000 & 0.938 & 8 & -16 \\ 1.000 & 1.016 & -2 & 4 \end{bmatrix} \begin{bmatrix} \mu \\ \nu \\ r_A \\ r_{C'} \end{bmatrix} = \begin{bmatrix} 0.719 \\ 0.734 \\ 0.813 \\ 0.891 \end{bmatrix} \tag{2.55}$$

Further reductions in weight are no longer possible, and Equation (2.55) defines the final solution to the linear program. This solution is

$$\mu = 0.708 \qquad \nu = 0.167$$

$$r_A = 0.00087 \qquad r_{C'} = 0.00370$$

The values of the moment capacities and hinge-rotation variables are therefore

$$\begin{aligned} M_p &= W_0L(1 - 0.708) = 0.292\,W_0L \\ M_p' &= W_0L(1 - 0.708 - 0.167) = 0.125\,W_0L \\ R_A &= 0.00087\,W_0L \\ R_{C'} &= 0.00370\,W_0L \end{aligned} \tag{2.56}$$

which corresponds to the point q_4 in the design space in Figure 2.11. Because of the linearity of the rigid–plastic contours, the moment capacity values above represent the final rigid–plastic design solution exactly. All of the linearization errors are absorbed in the values of the hinge-rotation variables.

Nevertheless, as will now be seen, it is useful to carry out a second design cycle. The calculations are begun by linearizing the response equations with respect to the solution obtained in the first cycle. A new linear program is formulated, and the optimization algorithm is executed a second time. In this case, the steps of the algorithm correspond to movements along the line $M_p = 2.33M_p'$ in the design space in Figure 2.11. A plastic hinge forms first at C′, then at A, and, finally, either at B or simultaneously at D and E. The corresponding proportional design and load history analysis are illustrated in Figure 2.13. The final solution refers to the point q_4 in Figure 2.11, and the results are

$$M_p = 0.292W_0L \qquad M_p' = 0.125W_0L$$

$$R_A = 0.00087W_0L \qquad R_{C'} = 0.00186W_0L$$

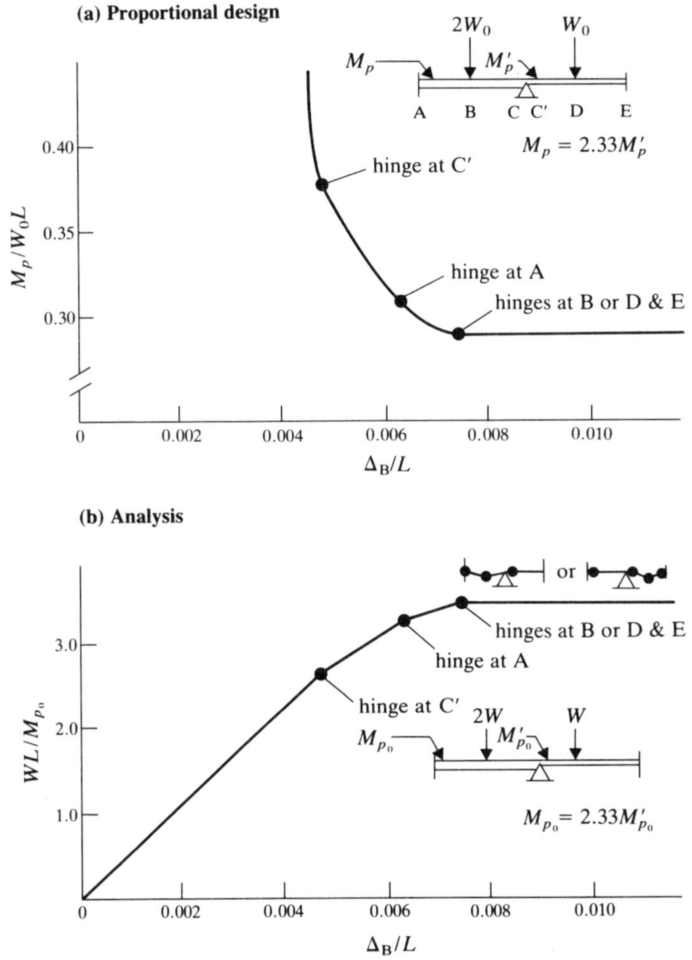

Figure 2.13 Analysis–design analogy for $M_p = 2.33M_p'$.

As expected, the values obtained for the moment capacities are the same as those determined in the previous cycle. The values obtained for the hinge rotation variables provide the final hinge rotations, which are

$$\phi_A = -\ 1.11 R_A/M_p = -1.11 \times 0.00087/0.292 = -0.0033 \text{ rad}$$

$$\phi_{C'} = -1.11 R_{C'}/M_p' = -1.11 \times 0.00186/0.125 = -0.0165 \text{ rad}$$

These are the hinge rotations immediately prior to the formation of the mechanism.

Reference

Reinschmidt, K. F., C. A. Cornell and J. F. Brotchie 1966. Interactive design and structural optimization. *J. Struct. Divn, ASCE* **89**(ST6), 281–318.

3

The proposed design approach

In the design of the continuous beam, an initial solution was required as a starting point, and the calculations were carried out in a series of design cycles. Each cycle involved linearizing the design equations, and solving the linear optimization problem which was generated.

For larger and more general design applications, the procedure is essentially the same, and is described in terms of a flowchart in Figure 3.1. As indicated in the flowchart, four distinct operations can be identified:

(a) design specification,
(b) initialization,
(c) formulation of the response functions,
(d) formulation of the linear program
(e) optimization.

Each in turn will now be described. For ease of presentation, attention will be confined to applications to two-dimensional steel frameworks.

3.1 Specification

In terms of the flowchart in Figure 3.1, the specification operation for the design of any structure refers to the setting up of the input data. As indicated in the figure, most of these data are the same as are required for an analysis of the structure. The remainder, which is referred to as the design data, is by far the smaller portion, and will now be described with reference to the cable-stayed bridge example in Figure 3.2.

The design data comprise the definition of the design variables, constraints and objective. As indicated in Figure 3.2, there are a variety of different variables and constraints. In the proposed procedure, all can be considered simultaneously in the search for the design solution.

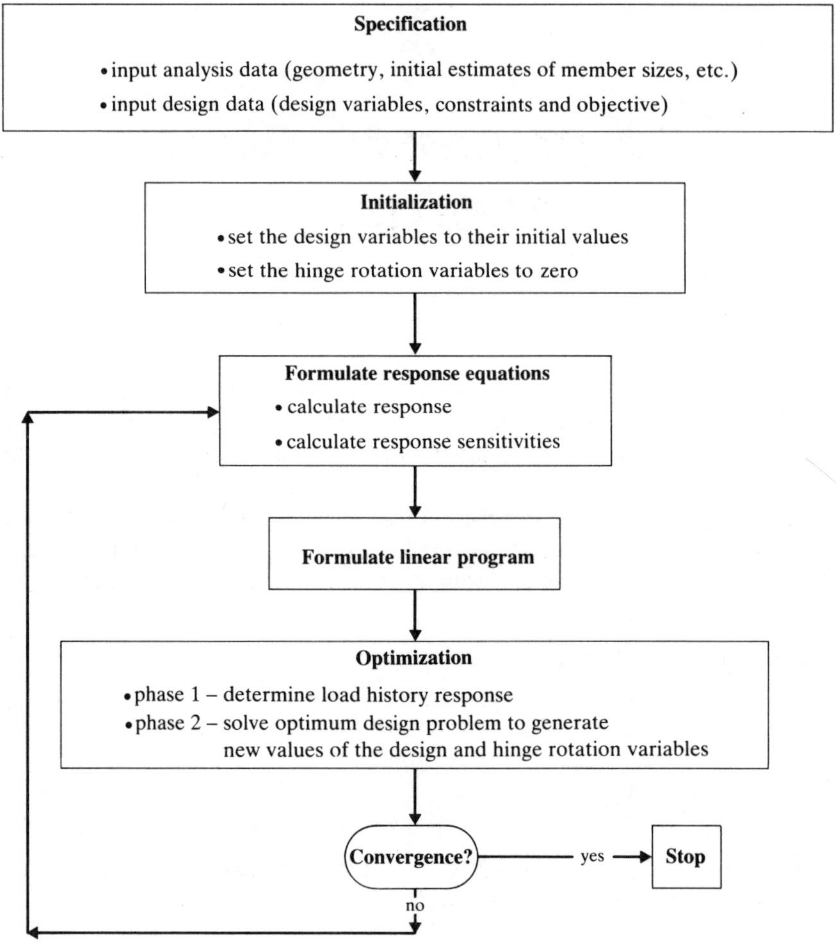

Figure 3.1 Simplified flowchart of the design approach.

3.1.1 Member size variables

For the case of the cable-stayed bridge, the member size variables are denoted in Figure 3.2a by the parameters v_1, v_2 and v_3. The three parameters refer to the cross-sectional areas of the girder, towers and cables of the bridge, respectively. Additional member size variables may be introduced whenever required. For example, if the designer wishes to explore solutions in which there is a change in cross section in the girder at its intersection with the towers, it is only necessary to specify an additional variable to refer to the cross-sectional areas of the girder in the side-spans. Alternatively, if the thickness and diameter of the tower section are to vary

(a) **Design variables**

(b) **Working load constraints**

$\Delta \leq \Delta_{w\,max}$
$P/A + M/Z \leq f_y$

(c) **Ultimate load constraints**

$\Delta \leq \Delta_{u\,max}$
$\phi \leq \phi_{max}$
$M \leq M_p$

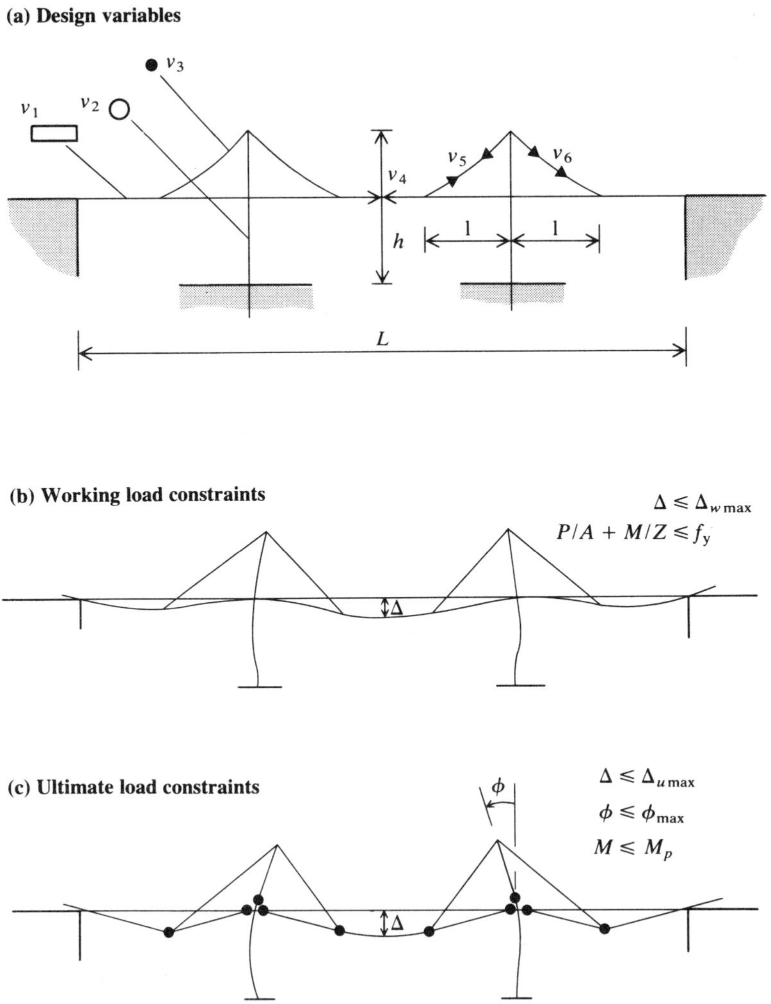

Figure 3.2 Design example: cable-stayed bridge.

independently of each other, it is only necessary to specify an additional variable such as the thickness itself. In general, therefore, variations can be considered not only in member sizes, but also in the shapes and proportions of the cross sections.

In design in practice, the shapes and proportions of cross sections are usually standardized, or are governed by practical and local instability considerations. One design variable is then sufficient to define any cross section, and relationships exist between the various section properties. Examples of these relationships for typical steel sections are shown in

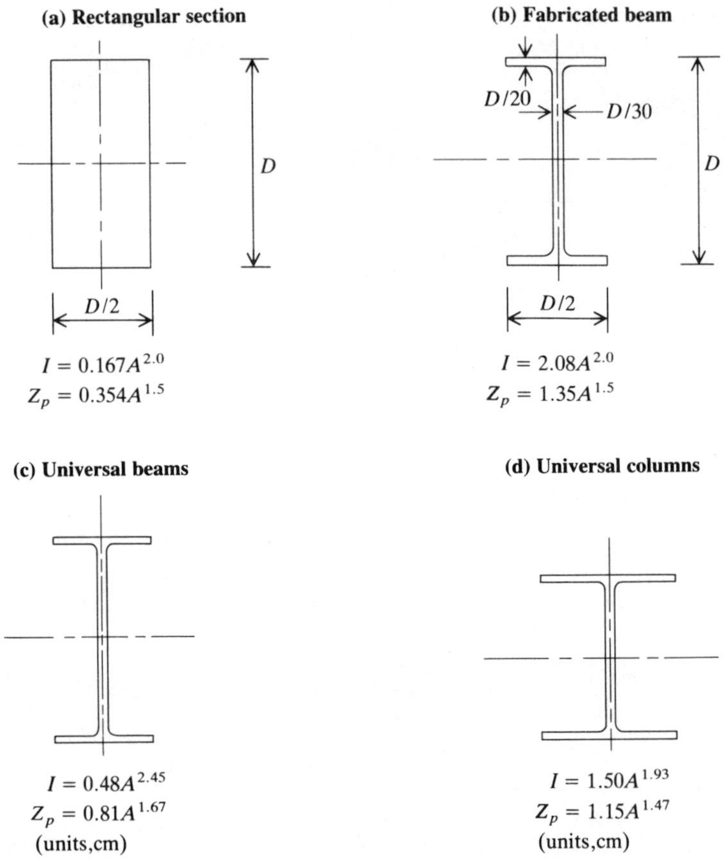

Figure 3.3 Section property relationships.

Figure 3.3. The relationships for the universal beams and columns are approximate, and were derived graphically as shown in Figure 3.4.

In any application of the design procedure, member sizes are chosen assuming that a continuous spectrum of universal section sizes is available. The procedure can be extended to determine an optimum solution with respect to a discrete list of sections, but the calculations involved are generally costly in computing time, and are not warranted in most practical applications.

In the first design cycle, the relationships between section properties are replaced by their linear approximations with respect to an initial design solution. The relationship between Z_p and A, for example, is replaced by the tangent at the starting point p_0 in Figure 3.5. The search for the optimum solution in the first cycle leads to a new point, p_1, on the same tangent. Since this new point generally does not lie on the original curve, the result of the first design cycle is more logically a choice between the

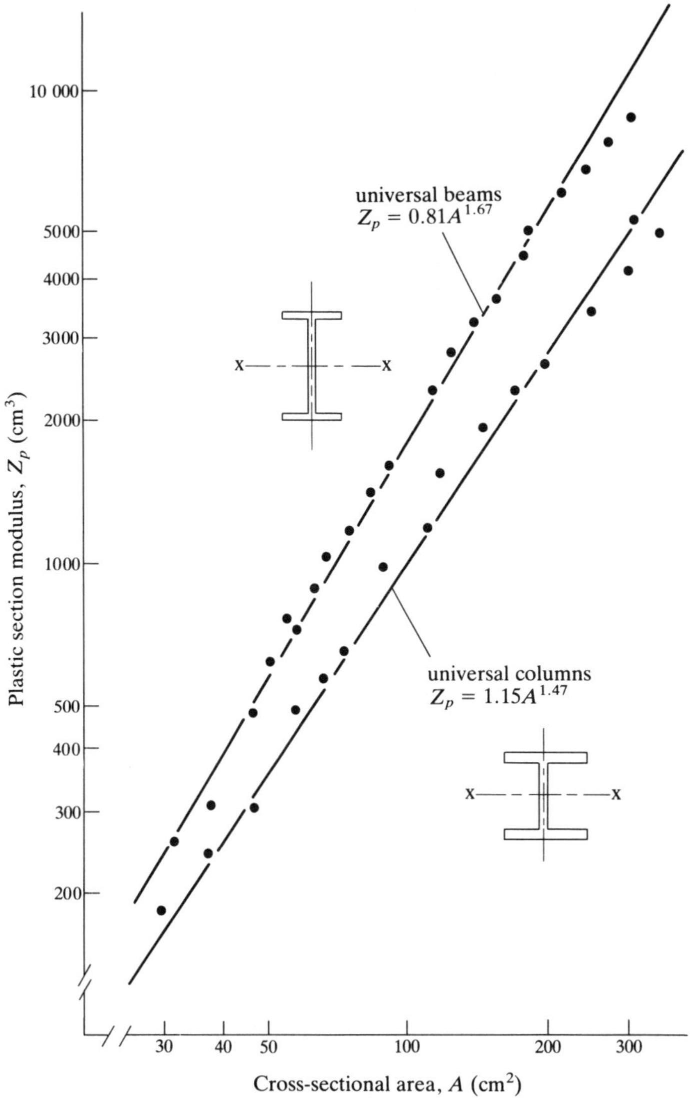

Figure 3.4 Section property relationships for universal beams and columns.

points p_2 and p_3 in the figure. In general, it is the former which is chosen since the section modulus, Z_p, is usually the dominant design variable. The latter is chosen in the case of truss members.

Whichever is adopted, the new point is used to replace p_0 as a basis for linearization in the second design cycle. The next and subsequent solutions are then modified in the same way, thereby improving convergence of the overall design procedure.

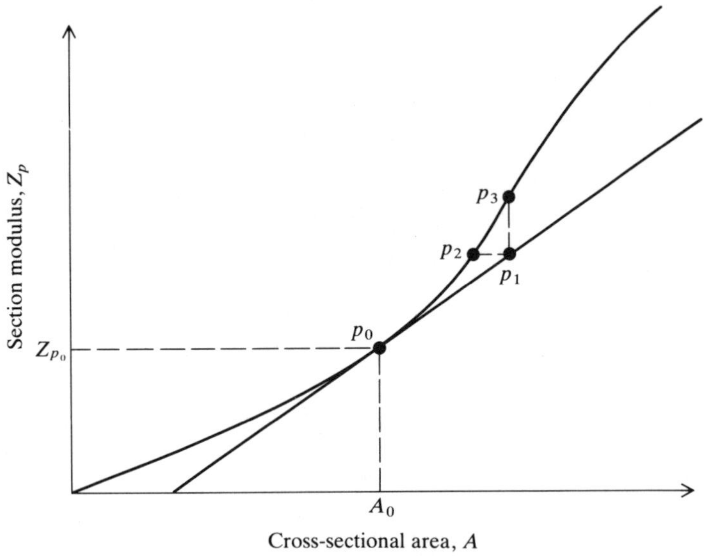

Figure 3.5 Linearization of a section property relationship.

3.1.2 Geometry and prestress variables

In the case of the cable-stayed bridge example in Figure 3.2a, only one geometry variable (v_4) is specified, and this refers to the height of the towers above the bridge deck. Other geometry variables can also be defined to explore, for example, variations in the inclination of the towers and changes in the locations of the cable supports. In the specification, each geometry variable is defined in terms of the relevant co-ordinates of the nodes of the structure. The variable v_4 is defined in terms of the vertical co-ordinates of the nodes at the top of the towers.

The levels of prestress in cables in a structure can also be specified as design variables. For the case of the bridge example in Figure 3.2a, there are two prestress variables, v_5 and v_6.

3.1.3 Objective function

As was discussed for the case of the continuous beam example, the objective function provides an indication of the merit of the design solution, and is assumed to correspond to the weight of the structure. The objective function for the cable-stayed bridge in Figure 3.2 is therefore

$$z = \rho v_1 L + 2\rho v_2 (h + v_4) + 4\rho v_3 \surd(l^2 + v_4^2) \tag{3.1}$$

where ρ is the density of steel.

In practice, the designer may wish to modify the function to make it represent design economy more accurately. For example, additional terms which are a function of the variables v_5 and v_6 may be included to incorporate the cost of prestressing. Alternatively, the coefficient of the last term may be modified to account for the higher unit costs of the high-strength steel cables.

3.1.4 *Yield constraints and hinge-rotation variables*

As in the case of the continuous beam example, yield constraints refer to the formation of plastic hinges in the structure, and their general form is

$$|M| \leqslant M_p \qquad |M| \leqslant 1.18M_p(1 - |P|/P_y) \tag{3.2}$$

where P and P_y denote, respectively, axial force and axial force capacity. The constraints above are discontinuous, and represent a total of six equivalent continuous constraints, which are illustrated in Figure 3.6, and which are as follows:

$$M \leqslant M_p \qquad -M \leqslant M_p \qquad \text{(3.3a, b)}$$

$$M \leqslant 1.18M_p(1 - P/P_y) \qquad -M \leqslant 1.18M_p(1 - P/P_y) \qquad \text{(3.3c, d)}$$

$$M \leqslant 1.18M_p(1 + P/P_y) \qquad -M \leqslant 1.18M_p(1 + P/P_y) \qquad \text{(3.3e, f)}$$

If all of these constraints are formulated for each load condition and each cross section in the structure, the resulting linear program would be unnecessarily large. For a particular load condition and cross section, only one dominates, and the constraint concerned does not generally change from one design cycle to the next. It is therefore often appropriate to formulate what will be referred to as the principal constraint, and, in some cases, one or two secondary constraints. The decision as to which of the six constraints in Equations (3.3) are principal and secondary is a function of the relevant response for the current design. For example, if the response for a particular load condition and cross section corresponds to a point in the area A in Figure 3.6, the principal constraint refers to the line a in the same figure. The two secondary constraints may then be simply the two adjoining lines, c and e. In any given application, the designer may specify the cross sections and load conditions in which secondary constraints are to be formulated.

Associated with each set of yield constraints is a specific plastic hinge-rotation variable. If the designer chooses not to specify that variable, plastic deformation is effectively avoided for the load condition and cross section concerned. If, on the other hand, the variable is specified, the value it

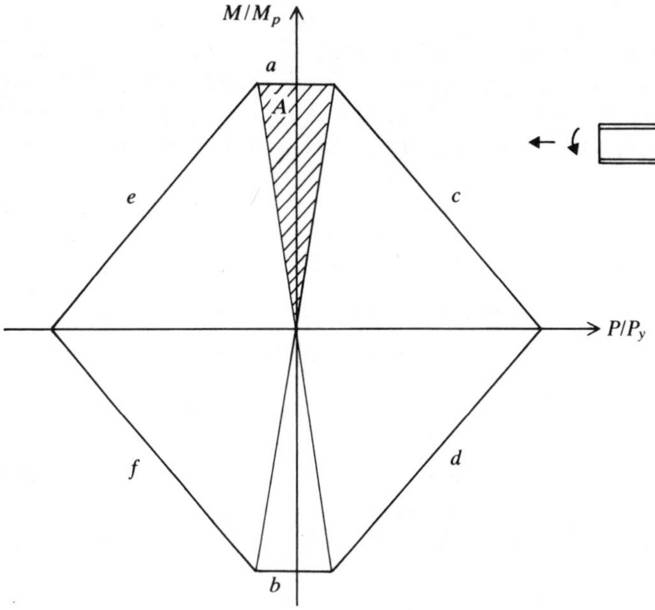

Figure 3.6 Yield constraints.

subsequently adopts is a measure of the extent to which elastic compatibility is violated at that section. In the extreme case that no hinge-rotation variables are specified, the calculations reduce to those of elastic design.

3.1.5 Additional constraints

As is shown in Figures 3.2b and c, the designer may wish to limit deflections and stresses at working load, and deflections and plastic hinge rotations at ultimate load. Limits on hinge rotations are referred to as ductility constraints, and they prevent the onset of a premature brittle or instability failure of the structure. Constraints may also be imposed on the design variables themselves. If the height of the bridge towers in Figure 3.2 is to be restricted because of aesthetic considerations, for example, the designer may wish to impose a constraint on the geometry variable v_4. If there is a minimum cable size to be used, the designer may wish to impose a constraint on the variable v_3. Alternatively, constraints on the design variables may be required for computational reasons. In highly non-linear problems, such as is generally the case when geometry variables and deflection constraints are involved, linearization errors can become excessive if the changes in the values of the variables from one design cycle to the next are too large. So-called move limits (Reinschmidt *et al.* 1966) may therefore be required to limit these changes during the optimization calculations.

3.2 Initialization

As is shown in the flowchart in Figure 3.1, specification is followed by initialization of the values of the variables. The independent design variables are set equal to the values defined in the input, and the hinge rotation variables are set equal to zero. The solution so defined forms the basis of the formulation in the first design cycle.

3.3 Formulation of the response equations

As will be recalled from the discussion of the continuous beam example, the response equations are formulated in terms of their linear approximations. In any one design cycle, these linear approximations are first-order Taylor expansions with respect to the current values of the design and hinge-rotation variables, and can be expressed in the form

$$\mathbf{X} = \mathbf{X}_0 + \sum_i \frac{\partial \mathbf{X}}{\partial v_i} \bar{v}_i + \sum_j \frac{\partial \mathbf{X}}{\partial R_j} \bar{R}_j \tag{3.4}$$

where

$\mathbf{X}$ is the vector of nodal deformation response,

$\mathbf{X}_0$ refers to the current design solution,

v_i is a design variable,

R_j is a hinge rotation variable,

$\bar{v}_i$ and $\bar{R}_j$ represent changes in the values of the variables and

$\partial\mathbf{X}/\partial v_i$ and $\partial\mathbf{X}/\partial R_j$ are the sensitivities of response to changes in the values of the variables.

A similar expression can be written for the so-called stress resultant response of the structure:

$$\mathbf{SR} = \mathbf{SR}_0 + \sum \frac{\partial \mathbf{SR}}{\partial v_i} \bar{v}_i + \sum \frac{\partial \mathbf{SR}}{\partial R_j} \bar{R}_j \tag{3.5}$$

where **SR** is the vector of stress resultants (i.e. axial forces, shear forces and bending moments in the members).

The formulation of response equations in any one design cycle therefore involves calculating the response for the current design, and the sensitivities of response to changes in both the design and hinge-rotation variables. This formulation will now be described in more detail.

3.3.1 Calculation of response

Response is calculated using the matrix stiffness method of analysis. The stiffness matrix is derived on the basis of an elastic structure, and the effects of yielding are considered by combining the loads with the effects of the hinge rotations. The relevant equations are derived from the equilibrium, compatibility and member stiffness relationships for the structure.

The equilibrium and compatibility relations are

$$\mathbf{W} = A\mathbf{SR} \qquad \mathbf{x} = A^{\mathrm{T}}\mathbf{X} \tag{3.6, 3.7}$$

where

- **W** and **X** are loads and deformations associated with the nodes of the structure,
- **SR** and **x** are stress resultants and deformations associated with members and
- A and A^{T} are the statics matrix and its transpose.

The two relationships are shown in Figures 3.7a and b for a typical member AB in the structure. The notation for the forces and deformations in the member are shown in Figure 3.8.

The member stiffness relationship is expressed in its general form as

$$\mathbf{SR} = S\mathbf{x} + \mathbf{SP} + S\boldsymbol{\phi} \tag{3.8}$$

where

- S is the member stiffness matrix,
- **SP** is the vector associated with loads on members and
- $\boldsymbol{\phi}$ is the vector of hinge rotations.

The solution for the stress resultant response may therefore be described as the superposition of the three separate solutions on the right-hand side of this equation. The first, $S\mathbf{x}$, is often referred to as the complementary solution, and is defined for a typical member in Figure 3.7c.

The second, **SP**, is referred to as the particular solution, and is defined for a uniformly distributed member load in Figure 3.9. The third, and last, solution, $S\boldsymbol{\phi}$, represents the contribution of the plastic hinge rotations. It is shown in Figure 3.10 for the case of a hinge rotation at end A of member AB. Contributions from hinge rotations at other locations, such as end B or anywhere along the member, can be defined similarly (Zeman 1975).

An alternative expression for **SR** can be obtained by combining Equations (3.7) and (3.8) to yield

$$\mathbf{SR} = SA^{\mathrm{T}}\mathbf{X} + \mathbf{Q} \tag{3.9}$$

(a) Equilibrium

$$\mathbf{W} = A\mathbf{S}\mathbf{R}$$

$$\begin{bmatrix} H_A \\ V_A \\ M_A \\ H_B \\ V_B \\ M_B \end{bmatrix} = \begin{bmatrix} \cos\alpha & -\sin\alpha & \\ \sin\alpha & \cos\alpha & \\ & & 1 \\ -\cos\alpha & \sin\alpha & \\ -\sin\alpha & -\cos\alpha & \\ & L & -1 \end{bmatrix} \begin{bmatrix} P \\ S \\ M \end{bmatrix}$$

(b) Compatibility

$$\mathbf{x} = A^{\mathrm{T}}\mathbf{X}$$

$$\begin{bmatrix} X \\ Y \\ \theta \end{bmatrix} = \begin{bmatrix} \cos\alpha & \sin\alpha & & -\cos\alpha & -\sin\alpha & \\ -\sin\alpha & \cos\alpha & & \sin\alpha & -\cos\alpha & L \\ & & 1 & & & -1 \end{bmatrix} \begin{bmatrix} X_A \\ Y_A \\ \theta_A \\ X_B \\ Y_B \\ \theta_B \end{bmatrix}$$

(c) Stiffness

$$\mathbf{S}\mathbf{R} = S\mathbf{x}$$

$$\begin{bmatrix} P \\ S \\ M \end{bmatrix} = \begin{bmatrix} EA/L & & \\ & 12EI/L^3 & 6EI/L^2 \\ & 6EI/L^2 & 4EI/L \end{bmatrix} \begin{bmatrix} X \\ Y \\ \theta \end{bmatrix}$$

Figure 3.7 Matrix analysis equations for member AB.

where

$$\mathbf{Q} = \mathbf{S}\mathbf{P} + S\boldsymbol{\phi}$$

For a specified set of loads and hinge rotations, therefore, the stress resultants are a function of the nodal deformation vector **X**. The latter can be determined by combining Equations (3.6) and (3.9) to yield

$$\mathbf{W} = ASA^{\mathrm{T}}\mathbf{X} + A\mathbf{Q}$$

i.e.

$$K\mathbf{X} = \mathbf{W} - A\mathbf{Q} \tag{3.10}$$

where $K = ASA^{\mathrm{T}}$ is the elastic frame stiffness matrix.

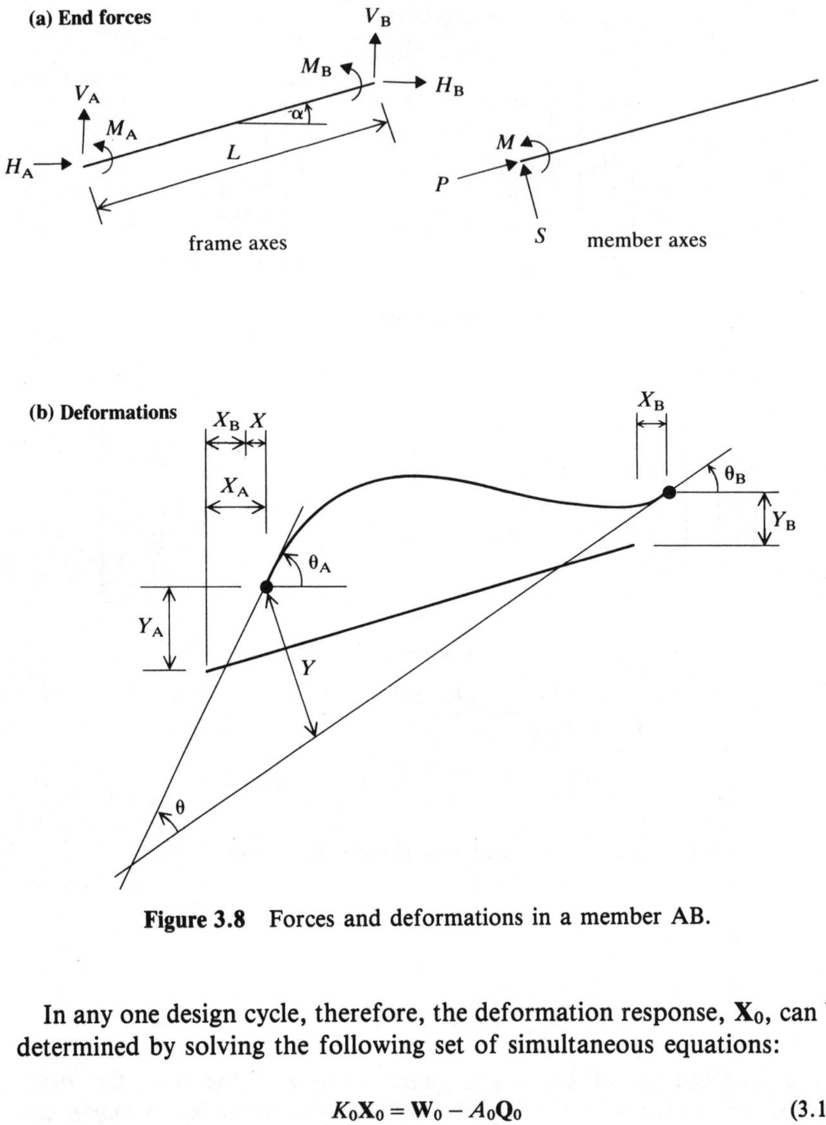

Figure 3.8 Forces and deformations in a member AB.

In any one design cycle, therefore, the deformation response, $\mathbf{X}_0$, can be determined by solving the following set of simultaneous equations:

$$K_0\mathbf{X}_0 = \mathbf{W}_0 - A_0\mathbf{Q}_0 \tag{3.11}$$

where the zero subscript refers to the current values of the design and hinge rotation variables.

The equations can be solved using the Choleski method for symmetric banded matrices. Three operations are involved, and they are, respectively, the reduction of the matrix into its lower and upper triangular forms, forward substitution and backward substitution.

Once the nodal deformations $\mathbf{X}_0$ are determined, they can be used to calculate the stress resultants $\mathbf{SR}_0$ at the member ends. The equation is the

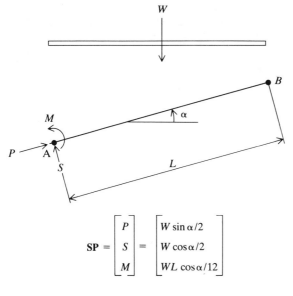

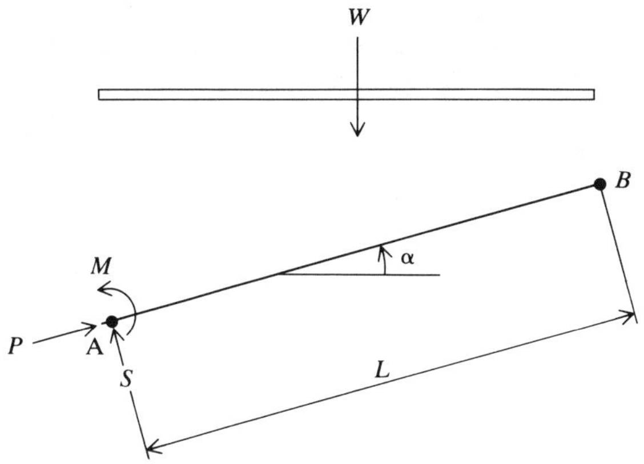

$$\mathbf{SP} = \begin{bmatrix} P \\ S \\ M \end{bmatrix} = \begin{bmatrix} W \sin\alpha/2 \\ W \cos\alpha/2 \\ WL \cos\alpha/12 \end{bmatrix}$$

(a) Member axes

$$A\ \mathbf{SP} = \begin{bmatrix} H_A \\ V_A \\ M_A \\ H_B \\ V_B \\ M_B \end{bmatrix} = \begin{bmatrix} 0 \\ W/2 \\ WL \cos\alpha/12 \\ 0 \\ W/2 \\ -WL \cos\alpha/12 \end{bmatrix}$$

(b) Frame axes

Figure 3.9 Particular solution for a distributed member load.

same as Equation (3.9) for the case of the current design, i.e.

$$\mathbf{SR}_0 = S_0 A_0^T \mathbf{X}_0 + \mathbf{Q}_0 \qquad (3.12)$$

Stress resultants along members can then be determined using equations of the form shown in Figure 3.11.

Once the response has been determined, the stress resultants can be combined at the nodes of the structure to ensure that they are in equilibrium with the external forces. The operation of combining the stress resultants at the nodes is equivalent to premultiplying the vector $\mathbf{SR}_0$ in Equation

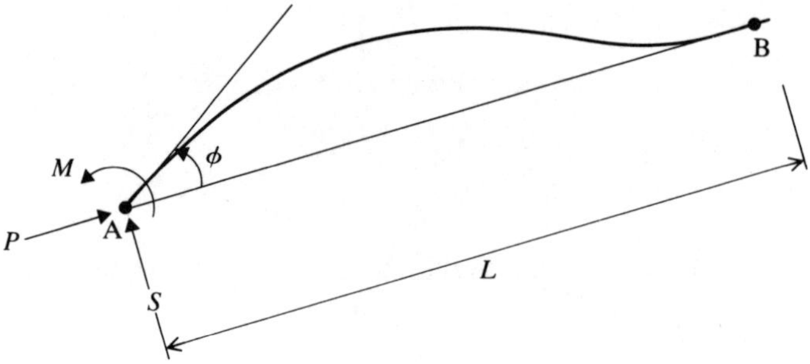

$$S\boldsymbol{\phi} = \begin{bmatrix} P \\ S \\ M \end{bmatrix} = \begin{bmatrix} EA/L & & \\ & 12EI/L^3 & 6EI/L^2 \\ & 6EI/L^2 & 4EI/L \end{bmatrix} \begin{bmatrix} 0 \\ 0 \\ \phi \end{bmatrix}$$

$$= \begin{bmatrix} 0 \\ 6R/L \\ 4R \end{bmatrix}$$

where $R = (EI/L)\,\phi$

Figure 3.10 The effect of an imposed hinge rotation: rotation at A of a member AB.

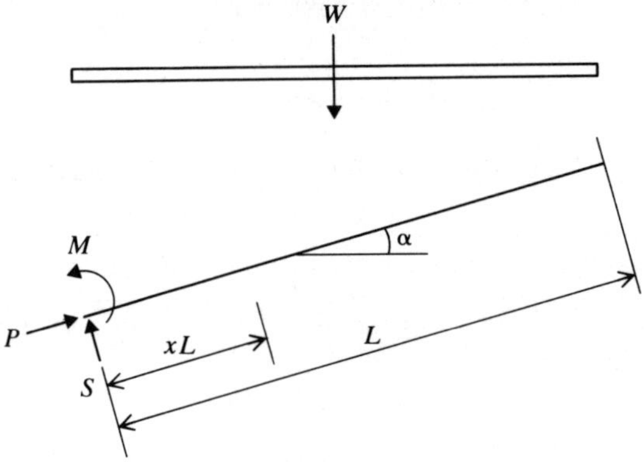

$$P_x = P - W(\sin\alpha)x$$
$$S_x = S - W(\cos\alpha)x$$
$$M_x = M - SLx + WL(\cos\alpha)x^2/2$$

Figure 3.11 Stress resultants along a member.

(3.12) by the statics matrix A_0, and leads to

$$\begin{aligned} A_0\mathbf{S}\mathbf{R}_0 &= A_0S_0A_0^{\mathrm{T}}\mathbf{X}_0 + A_0\mathbf{Q}_0 \\ &= K_0\mathbf{X}_0 + A_0\mathbf{Q}_0 \\ &= \mathbf{W}_0 \end{aligned}$$

The internal and external forces in the structure are therefore in equilibrium.

3.3.2 Response sensitivities

As is indicated by Equations (3.4) and (3.5), the formulation operation involves calculating not only the response for the current values of the variables, but also the sensitivities of response to changes in these values. Response sensitivities can be calculated using either of two procedures. The first is the more obvious procedure of independently incrementing the value of each variable in turn by an arbitrary small amount, and calculating the changes in response. This involves repetitive modifications to the stiffness matrix, and may therefore lead to excessive requirements in both computing time and storage.

A far more efficient procedure is one in which the response equations are differentiated analytically. This procedure will now be described for the calculation of sensitivities with respect to member size, geometry and hinge rotation variables. For convenience of presentation, the case of geometry variables will only be outlined briefly, and the case of prestress variables will be omitted. A more comprehensive treatment has been documented elsewhere (Zeman 1975).

MEMBER SIZE VARIABLES

As discussed in Section 3.1.1, a member size variable v is assumed to denote the cross-sectional area of one or more members in the structure. A change in the value of the variable results in a change in certain elements of the member stiffness matrix S. For small changes, this effect can be represented by the matrix

$$\bar{S} = \frac{\partial S}{\partial v}\,\bar{v}$$

where $\partial S/\partial v$ is a matrix, and is defined for a typical member in Figure 3.12, and $\bar{v}$ is a scalar representing the change in magnitude of the design variable. For design problems in which the loads themselves are a function of the member sizes (e.g. self weight, wave loads), changes also occur in the

$$S = \begin{bmatrix} EA/L & & \\ & 12EI/L^3 & 6EI/L^2 \\ & 6EI/L^2 & 4EI/L \end{bmatrix}$$

$$\frac{\partial S}{\partial v} = \begin{bmatrix} E/L & & \\ & 12EI'/L^3 & 6EI'/L^2 \\ & 6EI'/L^2 & 4EI'/L \end{bmatrix}$$

where $v = A$, $I' = \mathrm{d}I/\mathrm{d}A$

Figure 3.12 The member stiffness matrix and its sensitivity to changes in a cross-sectional area.

vector **SP** as follows:

$$\overline{\mathbf{SP}} = \frac{\partial \mathbf{SP}}{\partial v} \bar{v}$$

where $\partial \mathbf{SP}/\partial v$ is determined by differentiating the elements of the vector **SP** with respect to v.

Changes in the response of the structure can be written in a similar form:

$$\bar{\mathbf{X}} = \frac{\partial \mathbf{X}}{\partial v} \bar{v} \qquad \overline{\mathbf{SR}} = \frac{\partial \mathbf{SR}}{\partial v} \bar{v} \qquad (3.13, 14)$$

where $\partial \mathbf{X}/\partial v$ and $\partial \mathbf{SR}/\partial v$ are derived by differentiating the analysis equations (3.9) and (3.10). Equation (3.10) is

$$K\mathbf{X} = \mathbf{W} - A\mathbf{Q}$$

$$= \mathbf{W} - A(\mathbf{SP} + S\boldsymbol{\phi})$$

As will be seen in the section 'Hinge rotation variables', the hinge rotation variables are defined in such a manner that the product $S\boldsymbol{\phi}$ is independent of the member size variables. So, too, are the load vector **W** and the statics matrix A. Differentiation of the above equation with respect to a member size variable v therefore leads to

$$K_0 \frac{\partial \mathbf{X}}{\partial v} + \frac{\partial K}{\partial v} \mathbf{X}_0 = -A_0 \frac{\partial \mathbf{SP}}{\partial v}$$

i.e.

$$K_0 \frac{\partial \mathbf{X}}{\partial v} = -\frac{\partial K}{\partial v} \mathbf{X}_0 - A_0 \frac{\partial \mathbf{SP}}{\partial v}$$

$$= -A_0 \left(\frac{\partial S}{\partial v} A_0^{\mathrm{T}} \mathbf{X}_0 + \frac{\partial \mathbf{SP}}{\partial v} \right) \qquad (3.15)$$

where A_0, K_0 and $\mathbf{X}_0$ refer to the current design solution.

Equation (3.9) is

$$\mathbf{SR} = SA^{\mathrm{T}}\mathbf{X} + \mathbf{Q}$$

$$= SA^{\mathrm{T}}\mathbf{X} + \mathbf{SP} + S\boldsymbol{\phi}$$

and differentiation with respect to v leads to

$$\frac{\partial \mathbf{SR}}{\partial v} = S_0 A_0^{\mathrm{T}} \frac{\partial \mathbf{X}}{\partial v} + \frac{\partial S}{\partial v} A_0^{\mathrm{T}} \mathbf{X}_0 + \frac{\partial \mathbf{SP}}{\partial v} \tag{3.16}$$

The deformation response sensitivities, $\partial \mathbf{X}/\partial v$, in Equation (3.15) are the solution to a set of simultaneous equations in which the matrix of coefficients is again the elastic frame stiffness matrix K_0. If the Choleski method is used to solve the equations, the time-consuming first operation of matrix reduction need not be repeated, and the calculations reduce simply to a forward and backward substitution.

Once the deformation sensitivities have been computed, Equation (3.16) can be used to calculate the sensitivities of the stress resultants at the member ends. Sensitivities of stress resultants along members can then be deduced by differentiating equations of the form shown in Figure 3.11.

As in the case of the stress resultant response **SR**, the response sensitivities, $\partial \mathbf{SR}/\partial v$, can be premultiplied by the statics matrix A_0 to ensure that the solution satisfies equilibrium. Referring to both Equations (3.16) and (3.15),

$$\begin{aligned} A_0 \frac{\partial \mathbf{SR}}{\partial v} &= A_0 S_0 A_0^{\mathrm{T}} \frac{\partial \mathbf{X}}{\partial v} + A_0 \frac{\partial S}{\partial v} A_0^{\mathrm{T}} \mathbf{X}_0 + A_0 \frac{\partial \mathbf{SP}}{\partial v} \\ &= K_0 \frac{\partial \mathbf{X}}{\partial v} + A_0 \left(\frac{\partial S}{\partial v} A_0^{\mathrm{T}} \mathbf{X}_0 + \frac{\partial \mathbf{SP}}{\partial v} \right) \\ &= -A_0 \left(\frac{\partial S}{\partial v} A_0^{\mathrm{T}} \mathbf{X}_0 + \frac{\partial \mathbf{SP}}{\partial v} \right) + A_0 \left(\frac{\partial S}{\partial v} A_0^{\mathrm{T}} \mathbf{X}_0 + \frac{\partial \mathbf{SP}}{\partial v} \right) \\ &= \mathbf{0} \end{aligned}$$

where $\mathbf{0}$ is a vector containing zero elements only.

From Equation (3.14), it follows that

$$A_0 \overline{\mathbf{SR}} = A_0 \frac{\partial \mathbf{SR}}{\partial v} \bar{v} = \mathbf{0}$$

This equation reveals that no matter what changes are made to the new member sizes, there will be no additional resultant forces at the nodes of the structure. The linearization errors therefore affect only the compatibility equations, and not those of equilibrium.

GEOMETRY VARIABLES

A geometry variable is defined in terms of an X or Y co-ordinate of a specific joint in the structure. A change in the value of the variable generally results in modifications to the lengths and angles of inclination of the adjoining members, and these modifications can be expressed as

$$\bar{L} = \frac{\partial L}{\partial v}\bar{v} \qquad \bar{\alpha} = \frac{\partial \alpha}{\partial v}\bar{v}$$

where v denotes a geometry co-ordinate, and examples of results for $\partial L/\partial v$ and $\partial \alpha/\partial v$ are shown in Figure 3.13.

Modifications therefore also occur in the frame statics matrix, member stiffness matrix, member load vector and hinge rotation vector, and these modifications can be written as

$$\bar{A} = \frac{\partial A}{\partial v}\bar{v} \qquad \bar{S} = \frac{\partial S}{\partial v}\bar{v} \qquad \bar{\mathbf{Q}} = \frac{\partial \mathbf{Q}}{\partial v}\bar{v} \tag{3.17}$$

where the sensitivities $\partial A/\partial v$, $\partial S/\partial v$ and $\partial \mathbf{Q}/\partial v$ represent the sum of contributions from all members framing into the joint concerned. The calculation of the sensitivity $\partial A/\partial v$, for example, is illustrated in Figure 3.13.

As in the case of member size variables, the response sensitivities $\partial \mathbf{X}/\partial v$ and $\partial \mathbf{SR}/\partial v$ can be deduced by differentiating the analysis equations. For a geometry variable v, differentiation of Equations (3.10) and (3.9) yields

$$K_0\frac{\partial \mathbf{X}}{\partial v} = -\left(\frac{\partial K}{\partial v}\mathbf{X}_0 + \frac{\partial A}{\partial v}\mathbf{Q}_0 + A_0\frac{\partial \mathbf{Q}}{\partial v}\right) \tag{3.18}$$

$$\frac{\partial \mathbf{SR}}{\partial v} = G_0\frac{\partial \mathbf{X}}{\partial v} + \frac{\partial G}{\partial v}\mathbf{X}_0 + \frac{\partial \mathbf{Q}}{\partial v} \tag{3.19}$$

where

$$K_0 = A_0 S_0 A_0^{\mathrm{T}} = A_0 G_0$$

$$G_0 = S_0 A_0^{\mathrm{T}}$$

and

$$\partial K/\partial v = (\partial A/\partial v)G_0 + A_0(\partial G/\partial v)$$

$$\partial G/\partial v = (\partial S/\partial v)A_0^{\mathrm{T}} + S_0 \partial A^{\mathrm{T}}/\partial v$$

Deformation response sensitivities are determined by solving Equation

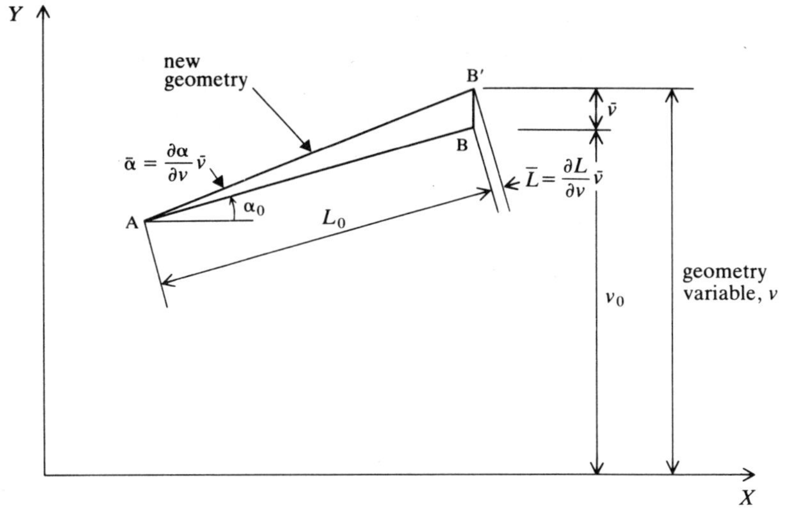

$$\frac{\partial A}{\partial v} = \begin{bmatrix} \partial(\cos\alpha)/\partial v & -\partial(\sin\alpha)/\partial v & 0 \\ \partial(\sin\alpha)/\partial v & \partial(\cos\alpha)/\partial v & 0 \\ 0 & 0 & 0 \\ -\partial(\cos\alpha)/\partial v & \partial(\sin\alpha)/\partial v & 0 \\ -\partial(\sin\alpha)/\partial v & -\partial(\cos\alpha)/\partial v & 0 \\ 0 & \partial L/\partial v & 0 \end{bmatrix}$$

where $\partial(\cos\alpha)/\partial v = -(\sin\alpha_0)\ \partial\alpha/\partial v = -(\sin\alpha_0)\ (\cos\alpha_0)/L_0$

$\partial(\sin\alpha)/\partial v = (\cos\alpha_0)\ \partial\alpha/\partial v = (\cos^2\alpha_0)/L_0$

$\partial L/\partial v = \sin\alpha_0$

Figure 3.13 Sensitivity of statics matrix to a change in the value of a geometry variable. Contribution of a member AB.

(3.18), and this is again an operation involving only forward and backward substitutions. The sensitivities of the stress resultants are then deduced using Equation (3.19).

HINGE ROTATION VARIABLES

As in the case of the continuous beam example, a convenient form for the hinge rotation variable is

$$R = (EI/L)\phi \qquad (3.20)$$

where ϕ is the hinge rotation and EI/L is the stiffness of the member to which the hinge rotation refers.

A change in the value of the hinge rotation variable R results in a change in the vector $\mathbf{Q}$, and is expressed as

$$\bar{\mathbf{Q}} = \frac{\partial \mathbf{Q}}{\partial R} \bar{R}$$

where the sensitivity, $\partial \mathbf{Q}/\partial R$, is obtained by differentiating the vector $\mathbf{Q}$ in Figure 3.10 with respect to R.

The equations for the response sensitivities can be derived by differentiating Equations (3.10) and (3.9) to yield

$$K_0 \frac{\partial \mathbf{X}}{\partial R} = -A_0 \frac{\partial \mathbf{Q}}{\partial R} \tag{3.21}$$

$$\frac{\partial \mathbf{SR}}{\partial R} = S_0 A_0^{\mathrm{T}} \frac{\partial \mathbf{X}}{\partial R} + \frac{\partial \mathbf{Q}}{\partial R} \tag{3.22}$$

Response is a linear function of the hinge-rotation variables, and the equations above are therefore free of linearization errors.

3.4 Formulation of the linear program

As shown in the flowchart in Figure 3.1, the next step in the design calculations is the formulation of the linear program. The program, which is simply the linear representation of the design problem, comprises an objective and a set of constraints. Both are functions of the variables in the design, and their formulation will now be described.

3.4.1 Objective function

It was discussed in Section 3.1.3 that, unless otherwise specified, the design objective is one of minimizing the weight of the structure. The objective function is therefore

$$z = \sum_m \rho_m A_m L_m \tag{3.23}$$

where ρ_m, A_m and L_m are the density, cross-sectional area and length of member, m, in the structure. The objective is therefore a function both of the member size and geometry variables in the design. If geometry variables are included, the function generally becomes non-linear, and its corresponding linearized form is

$$z = z_0 + \sum \eta_i \bar{v}_i \tag{3.24}$$

where

$$\eta_i = \partial z/\partial v_i$$

$$= \sum_m \left(\rho_m \frac{\partial A_m}{\partial v_i} L_m + \rho_m A_m \frac{\partial L_m}{\partial v_i} \right)$$

and where z_0 is the weight of the structure for the current design solution.

The variable v_i in the above equation may refer both to member sizes and to geometry. The summation term in the definition of η_i applies because more than one member may be associated with the same variable, v_i.

3.4.2 Design constraints

The design constraints are derived from the response functions discussed in Section 3.3, and are formulated in terms of both the design and hinge-rotation variables. Any constraint, whether it refers to a yield, ductility or serviceability criterion, can be expressed in the same general form, which will now be presented.

YIELD CONSTRAINTS

As will be recalled, the simplest form of yield constraint is

$$M \leqslant M_p \tag{3.25}$$

where M is a bending moment response and M_p is its limiting value. As was discussed in Section 3.3, response is generally a non-linear function of the design and hinge-rotation variables, and the linearized form for the bending moment response is

$$M = M_0 + \sum \alpha_i \bar{v}_i + \sum \beta_j \bar{R}_j \tag{3.26}$$

where

$$\bar{v}_i = v_i - v_{i0} \qquad \bar{R}_j = R_j - R_{j0}$$

$$\alpha_i = \partial M/\partial v_i \qquad \beta_j = \partial M/\partial R_j$$

and where the zero subscript refers to the values of the design variables, hinge rotations and bending moment response in the current design cycle. A more convenient form for Equation (3.26) can be derived by substituting $\bar{R}_j = R_j - R_{j0}$ to yield

$$M = M_0^{\text{el}} + \sum \alpha_i \bar{v}_i + \sum \beta_j R_j \tag{3.27}$$

where

$$M_0^{\text{el}} = M_0 - \sum \beta_j R_{j0}$$

The response M may therefore be described as the sum of three contributions corresponding to the three distinct terms in the above equation. The first term, M_0^{el}, refers to the current design, and is the value of M for zero values of the hinge-rotation variables. As was discussed in Section 3.3.2, response is a linear function of the hinge-rotation variables, and M_0^{el} is therefore simply the elastic response of the structure. The second term is the predicted change in response resulting from specific changes in the design, and the third term represents the effect of the hinge rotations.

The limiting value M_p is itself generally a function of one of the design variables, and the constraint equation (3.25) can therefore be written as

$$M_0^{\text{el}} + \sum \alpha_i \bar{v}_i + \sum \beta_j R_j \leqslant M_{p0} + \frac{\mathrm{d}M_p}{\mathrm{d}v_s}\, \bar{v}_s$$

i.e.

$$M_0^{\text{el}} + \sum \alpha_i' \bar{v}_i + \sum \beta_j R_j \leqslant M_{p0} \tag{3.28}$$

where

$$\alpha_i' = \begin{cases} \alpha_i - \dfrac{\mathrm{d}M_p}{\mathrm{d}v_s}\, \bar{v}_s, & i = s \\ \alpha_i, & i \neq s \end{cases}$$

If the variable v_s denotes a cross-sectional area A, then the value of $\mathrm{d}M_p/\mathrm{d}v_s$ is given by

$$\frac{\mathrm{d}M_p}{\mathrm{d}v_s} = f_y \frac{\mathrm{d}Z_p}{\mathrm{d}A} \tag{3.29}$$

where f_y is the yield strength of the member, Z_p is its plastic section modulus and $\mathrm{d}Z_p/\mathrm{d}A$ is derived from section property relationships, as in Figure 3.3.

Yield constraints involving axial loads as well as bending moments can be formulated in a similar manner. Equation (3.3c), for example, can be rewritten as

$$M' \leqslant M_p \tag{3.30}$$

where

$$M' = 0.85M + mP$$

and

$$m = M_p/P_y$$

For the current design, the value of M' is M_0', where

$$M_0' = 0.85M_0 + m_0P_0$$

and

$$m_0 = M_{p_0}/P_{y_0}$$

The linearized form for Equation (3.30) is its first-order Taylor expansion with respect to this current design, and is therefore

$$M_0' + \sum \alpha_i\bar{v}_i + \sum \beta_j\bar{R}_j \leqslant M_{p_0} + \frac{dM_p}{dv_s}\bar{v}_s \tag{3.31}$$

where

$$\alpha_i = 0.85\frac{\partial M}{\partial v_i} + m_0\frac{\partial P}{\partial v_i} + P_0\frac{\partial m}{\partial v_i}$$

$$\beta_j = 0.85\frac{\partial M}{\partial R_j} + m_0\frac{\partial P}{\partial R_j}$$

$$\frac{\partial m}{\partial v_i} = \begin{cases} \dfrac{1}{P_{y_0}}\left(\dfrac{dM_p}{dv_i} - m_0\dfrac{dP_y}{dv_i}\right), & i = s \\ 0, \quad i \neq s \end{cases}$$

and where dM_p/dv_i and dP_y/dv_i are defined by equations of the form of (3.29).

Equation (3.31) can therefore alternatively be written as

$$M_0'^{\text{el}} + \sum \alpha_i'\bar{v}_i + \sum \beta_jR_j \leqslant M_{p_0} \tag{3.32}$$

where

$$M_0'^{\text{el}} = M_0' - \sum \beta_jR_{j_0}$$

$$\alpha_i' = \begin{cases} \alpha_i - \dfrac{dM_p}{dv_i}, & i = s \\ \alpha_i, \quad i \neq s \end{cases}$$

which is of the same form as Equation (3.28), corresponding to the simple bending moment constraint.

DUCTILITY CONSTRAINTS

A ductility constraint is generally expressed as a maximum limit on the magnitude of a specific plastic hinge rotation, i.e.

$$\phi_j \leqslant \phi_{\max}$$

Expressed in terms of the corresponding hinge-rotation variable R_j, the constraint is

$$R_j \leqslant \frac{EI}{L}\phi_{\max} \tag{3.33}$$

where

$$R_j = \frac{EI}{L}\phi_j$$

As in the case of the yield criteria, the above constraint is represented in the linear program by its first-order Taylor expansion. The form of the expansion is a function of the nature of the relationship between the stiffness (EI/L) and the design variables, and can be represented as

$$R_j \leqslant \frac{EI_0}{L_0}\phi_{\max} + \frac{E}{L_0}\phi_{\max}\frac{\mathrm{d}I}{\mathrm{d}v_s}\bar{v}_s - \frac{EI_0}{L_0^2}\phi_{\max}\frac{\mathrm{d}L}{\mathrm{d}v_m}\bar{v}_m$$

i.e.

$$\sum \alpha_i\bar{v}_i + R_j \leqslant \frac{EI_0}{L_0}\phi_{\max} \tag{3.34}$$

where

$$\alpha_i = \begin{cases} \dfrac{-E}{L_0}\phi_{\max}\dfrac{\mathrm{d}I}{\mathrm{d}v_i}, & i = s \\ \dfrac{EI_0}{L_0^2}\phi_{\max}\dfrac{\mathrm{d}L}{\mathrm{d}v_m}, & i = m \\ 0, \quad i \neq s, m \end{cases}$$

and where v_s and v_m refer to specific member size and geometry variables, respectively.

SERVICEABILITY CONSTRAINTS

Serviceability constraints are defined at working load, and are generally limitations on the magnitudes of deflections, i.e.

$$\Delta \leqslant \Delta_{\max} \tag{3.35}$$

Assuming that there is no plastic deformation at working load, the linearized response function for Δ is

$$\Delta = \Delta_0 + \sum \alpha_i \bar{v}_i \tag{3.36}$$

where

$$\alpha_i = \partial\Delta/\partial v_i \qquad \Delta_0 = \Delta_0^{\text{el}}$$

Substitution of Equation (3.36) into Equation (3.35) leads to

$$\Delta_0 + \sum \alpha_i \bar{v}_i \leqslant \Delta_{\max} \tag{3.37}$$

Deflection is a measure of the flexibility of the structure, and is likely to be a highly non-linear function of the member size variables. The linear approximation expressed in Equation (3.37) may consequently lead to considerable linearization errors in the subsequent optimization. It is expected that a more efficient form may be derived by basing the linearization on stiffness rather than on flexibility. The stiffness version of the constraint equation (3.35) is simply its inverse, i.e.

$$\frac{1}{\Delta} \geqslant \frac{1}{\Delta_{\max}} \tag{3.38}$$

The first-order Taylor expansion of this equation differs from Equation (3.37), and can be derived as follows:

$$\frac{1}{\Delta_0} - \frac{1}{(\Delta_0)^2} \sum \alpha_i \bar{v}_i \geqslant \frac{1}{\Delta_{\max}}$$

Multiplication of both sides of this equation by the product $\Delta_0 \Delta_{\max}$ leads to

$$\Delta_{\max} - \frac{\Delta_{\max}}{\Delta_0} \sum \alpha_i \bar{v}_i \geqslant \Delta_0$$

i.e.

$$\Delta_0 + \sum \alpha_i' \bar{v}_i \leqslant \Delta_{\max} \tag{3.39}$$

where

$$\alpha_i' = \frac{\Delta_{\max}}{\Delta_0} \alpha_i$$

This is the form which is adopted in the design calculations.

3.5 Optimization

Once the linear program has been formulated, it is solved using the linear programming algorithm. In the case of the continuous beam example, as was discussed in Section 2.4.2, the algorithm was executed in two phases, and load history response was determined as a by-product of the optimization. This feature is attractive not only because of its obvious economic advantages, but also, as will be seen, because it allows consideration in design of load history phenomena such as hinge freezing.

As will be recalled, the load history response for the continuous beam was computed by making use of the analysis–design analogy discussed in Section 2.2. The analogy is strictly only valid, however, for design problems in which, firstly, cross-sectional properties A, I and M_p vary in linear proportion to each other and, secondly, the sizes of all the members throughout the structure are to be varied. Clearly, this may not always be the case.

In a general design, load history response can still be determined as a by-product of the optimization calculations, but the procedure needs to be modified, as will now be discussed.

3.5.1 *Initial solution*

A design problem subject to simple yield criteria can be expressed in terms of the following linear program:

minimize the objective function z, where

$$z = z_0 + \sum \eta_i \bar{v}_i$$

subject to constraints of the form

$$M_{0k}^{\text{el}} + \sum \alpha_{ki}' \bar{v}_i + \sum \beta_{kj} R_j \leqslant M_{p_{0k}}, \qquad k = 1,2,\ldots \tag{3.40}$$

where for a particular cross section and load condition, M_{0k}^{el} is the elastic bending moment response to the factored loads and $M_{p_{0k}}$ is the magnitude of the moment capacity for the current design. Design problems subject to constraints other than simple yield criteria can be expressed in a similar form using constraint equations (3.32), (3.34) and (3.39).

The corresponding analysis problem is one in which the values of all the design variables $\bar{v}_i$ are equal to zero, and variations are considered in the loading on the structure. Assuming these variations correspond to a proportional load history, the analysis problem can be expressed as follows:

maximize γ subject to constraints of the form

$$\gamma M_{0k}^{\mathrm{el}} + \sum \beta_{kj} R_j \leqslant M_{p0k} \tag{3.41}$$

where γ may be referred to as a normalized load factor variable, which for the continuous beam example corresponds to the ratio W/W_0. For a more general structure under any number of load conditions, it corresponds to

$$\gamma = \lambda_\tau / \lambda_{0\tau}, \qquad \tau = 1, 2, \ldots \tag{3.42}$$

where the subscript τ refers to a particular load condition, $\lambda_{0\tau}$ is the load factor for which the structure is to be designed and λ_τ is the load factor at any stage of loading. Serviceability criteria are associated with a separate load condition for which $\lambda_{0\tau}$ is equal to unity.

Load history response data can be determined as a by-product of the optimization calculations by combining the two linear programs as follows:

minimize $z = z_0' + \sum \eta_i \bar{v}_i - N\gamma$ subject to the following constraints:

$$\gamma \leqslant 1$$

and

$$\gamma M_{0k}^{\mathrm{el}} + \sum \alpha_{ki} \bar{v}_i + \sum \beta_{kj} R_j \leqslant M_{p0k}, \qquad k = 1, 2, \ldots \tag{3.43}$$

where

$$z_0' = z_0 + N$$

and where N is a very large positive number.

In terms of slack variables, the linear program is written as

minimize $z = z_0' + \sum \eta_i \bar{v}_i - N\gamma$ subject to the following:

$$\gamma + S_0 = 1$$

and

$$\gamma M_{0k}^{\mathrm{el}} + \sum \alpha_{ki} \bar{v}_i + \sum \beta_{kj} R_j + S_k = M_{p0k}$$

$$S_0 \geqslant 0 \qquad S_k \geqslant 0 \qquad k = 1, 2, \ldots \tag{3.44}$$

where the slack variable S_k may refer not just to yield constraints, but also to ductility, serviceability and other constraints.

The starting point for the linear programming calculations is the initial solution

$$\gamma = 0 \qquad \bar{v}_i = 0, \qquad i = 1,2, \ldots$$

$$R_j = 0, \qquad j = 1,2, \ldots$$

$$S_0 = 1 \qquad S_k = M_{pok}, \qquad k = 1,2, \ldots$$

3.5.2 *Load history phase*

The first phase is one in which the values of $\bar{v}_i$ are maintained at zero. The linear program reduces to one of maximizing the load factor variable without exceeding its design limit of unity, and without violating any of the design constraints. The yield constraints reduce to the same form as those of the corresponding analysis linear program in Equation (3.41), and the calculations reduce to those of an elastic–plastic analysis for the current design.

The sequence of calculations is essentially the same as that discussed for the continuous beam in the section 'Application of linear programming' in Section 2.1.2. The first iteration is an interchange between the load factor variable γ and one of the slack variables, S_k. The variable γ is increased from its initial value of zero, while the variable S_k reduces to zero. If S_k refers to a yield constraint, then the first iteration corresponds to the development of the first plastic hinge. The second iteration is then generally an interchange between a corresponding hinge-rotation variable R_j and a second slack variable, and the procedure is continued as each successive plastic hinge develops.

A plastic hinge is only permitted to rotate after the corresponding cross section has yielded. Consequently, the value of a hinge-rotation variable R_j is permitted to increase from zero only after the value of the corresponding slack variable S_k has reduced to zero. As is shown in Figure 3.14, a hinge rotation is positive or negative depending on the sign of the bending moment at the section. Once a section has yielded, therefore, the sign of the corresponding hinge-rotation variable is known even while its value is still zero. If that sign is positive, the following variable transformation is performed before continuing the calculations:

$$R_j = -R_j \tag{3.45}$$

In this way, the hinge-rotation variables in the linear program are forced to adopt positive values only. As was discussed in Section 2.1.2, this is a general requirement of the linear programming algorithm.

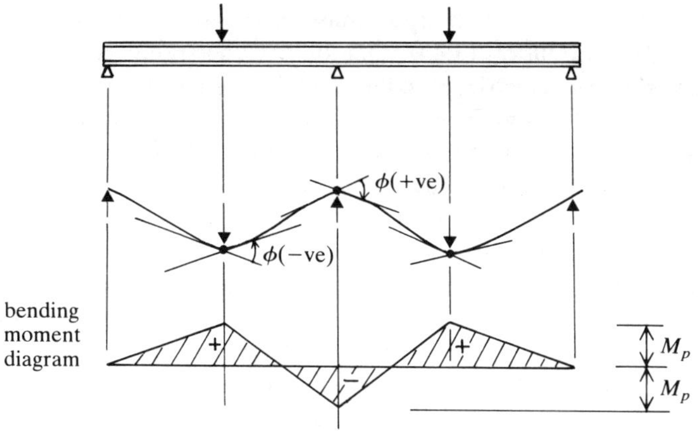

Figure 3.14 Sign convention for plastic hinge rotations.

Once the magnitude of a hinge rotation starts to increase, it generally continues to increase until the maximum loads are reached. If at any stage a hinge-rotation variable R_j is tending to decrease in value, the effects of hinge freezing need to be considered. Such effects can be considered using the variable transformation

$$Q_j = R_j - R_{j1} \tag{3.46}$$

where Q_j is the new hinge-rotation variable and R_{j1} is the current value of R_j. Because the variable Q_j is restricted to positive values, the variable R_j is subsequently frozen at its value of R_{j1}. The new variable therefore adopts a value of zero, and generally retains that value in the next iteration. The corresponding slack variable is then permitted to adopt positive values again, thus allowing the corresponding cross section to regain its elasticity.

At no stage in the calculations is a corresponding set of slack and hinge-rotation variables permitted to adopt non-zero values simultaneously. The calculations are continued until γ can no longer be increased without violating a yield, deflection or ductility constraint.

3.5.3 Design search phase

The second phase of the linear programming calculations is one in which the values of the design variables $\bar{v}_i$ are allowed to vary freely. Both negative and positive values are permitted, and the following variable transformation is therefore warranted:

$$\bar{v}_i = p_i - p_0, \qquad i = 1, 2, \ldots \tag{3.47}$$

where $p_i \geqslant 0$ and $p_0 \geqslant 0$ are the new variables in the program.

The objective of this second phase is to solve the linear program represented by Equation (3.43). One of the constraints in the program is that the variable γ is restricted to lie below its design value of unity. Because the variable has been assigned a large negative coefficient in the objective function, it will always reach its design value. The calculations therefore reduce to those corresponding to Equation (3.40), and the result is the required minimum weight solution for the design cycle.

The calculation procedure is similar to that in the first phase. The values of a hinge rotation and corresponding slack variable should never be non-zero simultaneously, and the variable transformation of Equation (3.45) is required whenever hinge rotations are negative. In contrast to the first phase, the values of the hinge-rotation variables are permitted to decrease at any stage in the calculations. An iteration can involve an interchange not only between a hinge-rotation and slack variable, but also between one hinge-rotation variable and another, or between one slack variable and another. The iterations are continued until the minimum weight solution is determined.

References

Reinschmidt, K. F., C. A. Cornell and J. F. Brotchie 1966. Iterative design and structural optimization. *J. Struct. Divn, ASCE* **89** (ST6), 281–318.

Zeman, P. 1975. *Optimum elastic–plastic design of framed structures, Vols. I and II.* Ph.D. thesis, University of Sydney, Australia.

4

Design examples

The proposed design method has been implemented using a computer program known as SODA (System for Optimum Design and Analysis). Two examples of its application will now be described. The examples are provided for demonstration only. Efforts were made to make them as practical as possible without complicating the treatment unnecessarily.

4.1 Low-rise industrial frame

The first example is the design of the two-bay pitched roof portal frame shown in Figure 4.1. The design problem was one of choosing, from standard universal beam sections, section sizes for the frame members. As illustrated in the figure, there were three distinct sizes, B1, B2 and B3, which refer to the outer columns, the inner column and the four rafters, respectively.

The yield stress and Young's modulus were assumed to be $f_y = 240\,\text{MPa}$ and $E = 210\,\text{GPa}$. The effect of axial force on moment capacity was neglected for this example.

As is shown in Figure 4.1, the ends of the rafters were reinforced by haunches. These haunches were designed to remain elastic under the imposed loads, and their details were determined after SODA was executed. Their effect on overall stiffness was included in the computer model in the form of the rigid arms shown in Figure 4.1b.

As is shown in Figure 4.2, the frame was subjected to two load conditions as follows:

(a) LC1 – dead + live + crane loads and
(b) LC2 – dead + crane + wind loads.

The frame was required to withstand the first condition at a load factor of 1.67, and the second at a load factor of 1.25. Dead and live load were assumed to be equal, and their combined effect was a downward load of 3 kN/m on the rafters, distributed uniformly. The crane loads were represented by eccentric concentrated vertical and horizontal forces on the columns, and the wind loads were represented by uniformly distributed

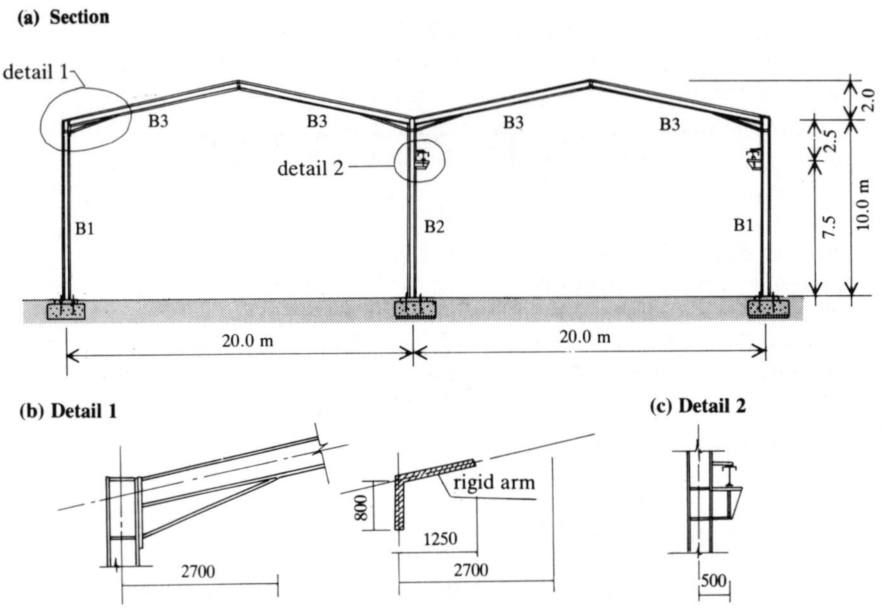

Figure 4.1 Low-rise industrial frame example.

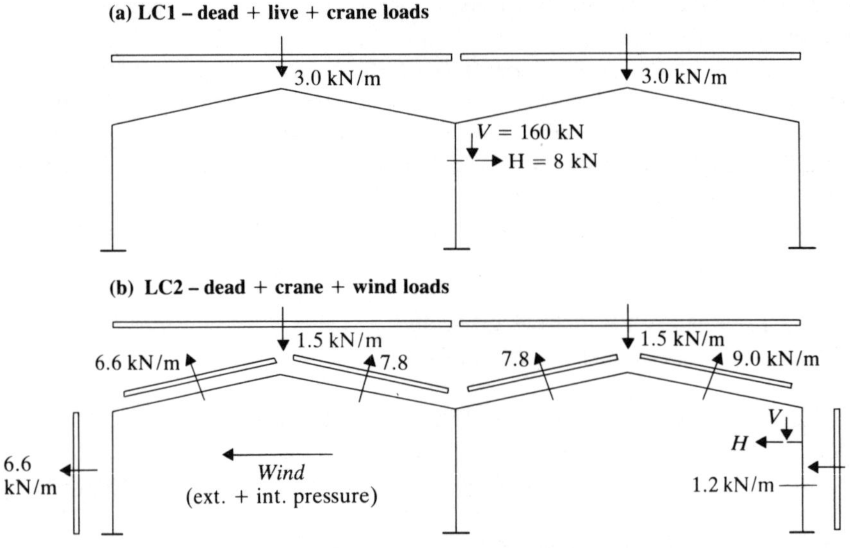

Figure 4.2 Load conditions.

loads applied normal to the face of the relevant members. The magnitudes of the wind loads were determined by combining the external pressure and suction on the frame with a constant internal pressure.

Two separate designs were undertaken, and each will now be described.

4.1.1 Rigid–plastic design

The first was a rigid–plastic design of the frame. The structure was simply required not to fail under the imposed load conditions, LC1 and LC2. The results for the design are shown in Figure 4.3a, together with the bending moment response for LC2 (dead + crane + wind loads) at the prescribed load factor, $\lambda = 1.25$. The load history response determined from the calculations is represented by the lower curve in Figure 4.4. As shown in the figure, four plastic hinges occurred in the rafters and two occurred in the windward column. The last two hinges formed simultaneously, and resulted in rigid–plastic collapse of the frame at the prescribed load factor.

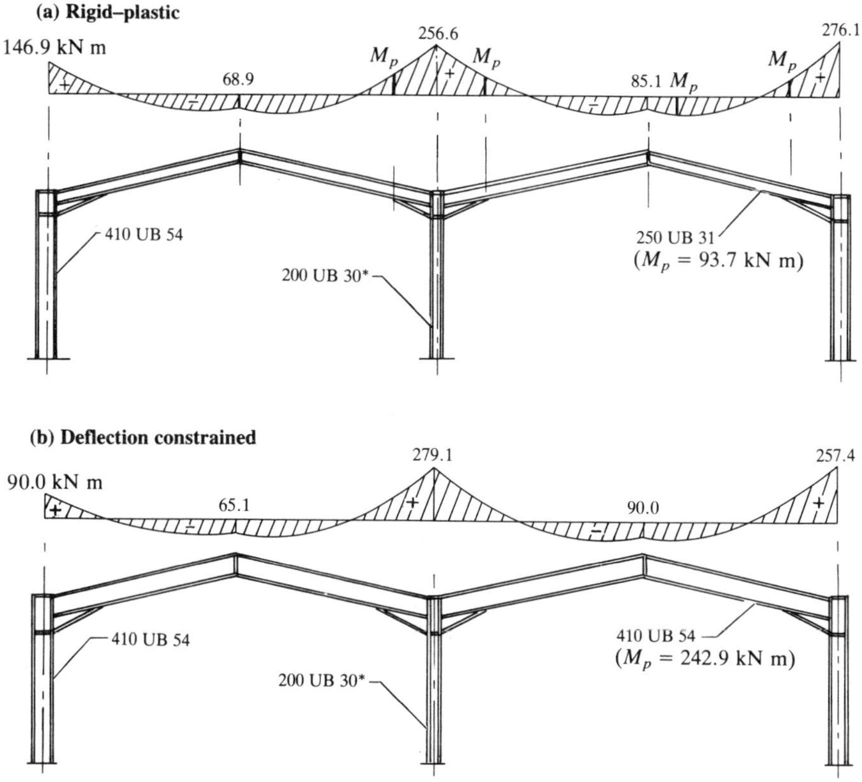

Figure 4.3 Design solutions: bending moment diagrams for LC2 ($\lambda = 1.25$). Note that the example is for demonstration only, and no allowance has been made for the effects of buckling.

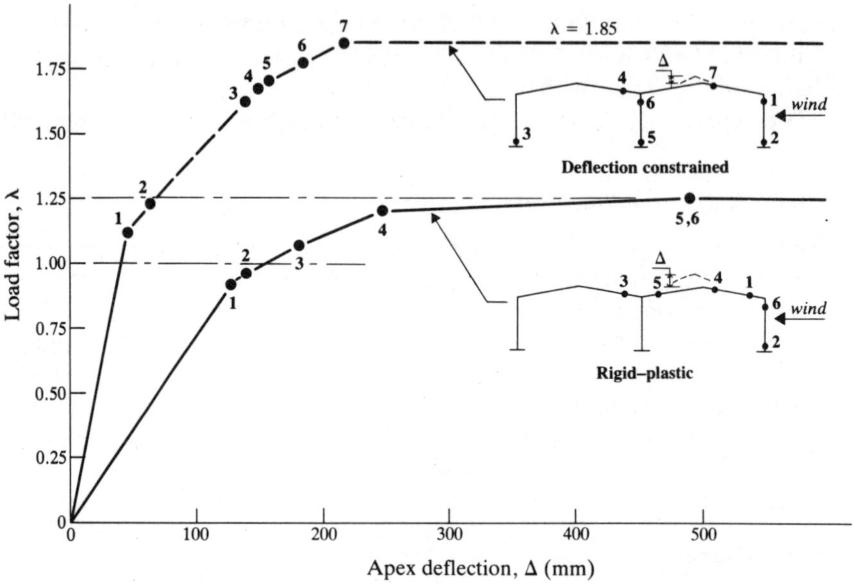

Figure 4.4 Load deflection diagrams for load condition LC2.

The frame also experienced failure at the prescribed load factor (λ = 1.67) under load condition LC1 (dead + live + crane). The failure in this case was a local one, and was characterized by plastic hinges only in the central column. Two hinges formed, one on either side of the crane support.

In terms of computational effort, only one design cycle was required to determine the optimum rigid–plastic solution, and the total computing cost was only A$10 (35 s of CPU time on a CYBER 171 computer). The results of the calculations included not only the design solution, but also load history, and deflection and stress response for both load conditions.

4.1.2 Deflection constrained design

The rigid–plastic solution is clearly a very economical one, but is not necessarily satisfactory, particularly with regards to response at working load, $\lambda = 1.0$. As is shown in Figure 4.4, plastic hinges form even before working load level is reached, and deflections are excessive.

A second design of the frame was therefore carried out to demonstrate the application of the method to design problems involving working load deflection constraints. The constraint which was imposed was

$$\Delta \leqslant L/400 = 50\,\mathrm{mm}$$

where Δ is the maximum vertical apex deflection under dead and wind load, and L is the rafter span. The design problem was therefore one of determining an optimum set of section sizes such that this constraint could be satisfied. At the same time, the frame was still required to withstand the two load conditions, LC1 and LC2, at the nominated load factors $\lambda = 1.67$ and $\lambda = 1.25$, respectively.

The results are shown in Figure 4.3b and the upper curve in Figure 4.4. The column sections were found to be the same as those for the rigid–plastic design, and there was a large increase in the size of the rafters. This increase led to a change in the total weight of the frame of 35%. As is shown in Figure 4.3, it also led to substantial increases in rafter bending moment response, yet no bending moment outside the haunches was large enough to exceed the section capacity ($M_p = 242.9$ kN m). As is shown in Figure 4.4, only two plastic hinges occurred below the prescribed load factor of $\lambda = 1.25$, and both formed in the windward column. The frame therefore had a reserve of strength well above $\lambda = 1.25$. The extent of this reserve is a function of the strength of the haunches. If the haunches are designed to remain elastic, the response is as shown in Figure 4.4, and collapse does not occur until $\lambda = 1.85$.

In terms of computational effort, the design problem was a non-linear one, and required the use of the move limits referred to in Section 3.1.5. The number of design cycles required for convergence varied, and was a function both of the initial design solution and of the size of the move limits.

A useful strategy in solving a deflection constrained design problem such as this is first to determine the corresponding rigid–plastic solution, and then to use that as a starting point for subsequent calculations. The required solution is then generally obtained relatively rapidly. Further reductions in computer costs can be achieved by replacing the yield constraints (and hinge-rotation variables) with lower bound limits on the design variables themselves. By restricting section sizes to values exceeding those determined in the rigid–plastic design, the safety of the structure is ensured. Although the final solution is not necessarily optimal, it is generally a very satisfactory one.

4.2 Multistorey frame in seismic zone

The second example was the design of the 12-storey, three-bay frame shown in Figure 4.5. The design problem was one of choosing section sizes for the frame members from standard universal beam and column sections. As is illustrated in the figure, there was a total of 12 distinct sizes, B1 to B6 and

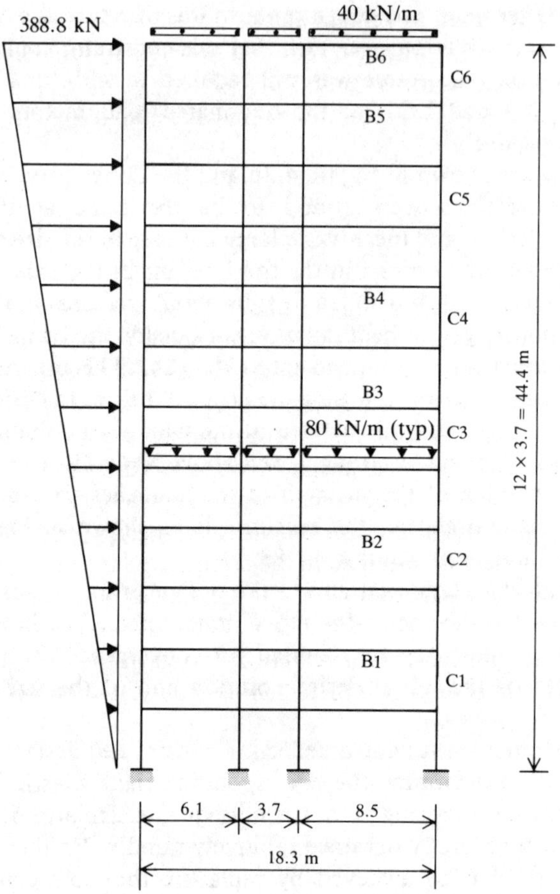

Figure 4.5 Twelve-storey frame example: vertical and earthquake loads.

C1 to C6. Beam and column sections were permitted to change at every two storeys, except for the beams near roof level.

The values of yield stress and Young's modulus were assumed to be $f_y = 360\,\text{MPa}$ and $E = 210\,\text{GPa}$. The effect of axial force on moment capacity was considered. The relevant equations are described in Section 3.1.4.

As is shown in Figure 4.5, the frame was subjected to a combination of vertical (dead + live) and earthquake loading. The latter was applied in the form of a triangular distribution, and corresponded to a horizontal acceleration of 0.15 **g**. Two load conditions were considered:

(a) LC1 – vertical loads alone and
(b) LC2 – vertical + earthquake loads.

The frame was required to remain elastic under the first load condition at a load factor of 1.70. It was also required to withstand the second without collapsing at a load factor of unity.

Two separate designs were undertaken, and they differ in terms of the design requirements for LC2.

4.2.1 Rigid–plastic design

In the first design, plastic hinges under LC2 were permitted to form anywhere in the structure. The results are shown in the form of required section moduli in Table 4.1 and the lower curve in Figure 4.6. As indicated by the curve, the frame reaches rigid–plastic collapse under LC2 at the prescribed unit load factor. The corresponding sway at roof level was 235 cm, and the collapse mechanism involved a sidesway of 11 storeys.

The absence of plastic hinges near roof level was attributed to the fact that the selection of the beam size, B6, was governed by load condition LC1 (vertical loads alone). As specified in the criteria, the frame remained elastic under LC1 for the factored loads. Plastic hinges were on the verge of forming at two locations, both in beams near roof level.

The total computer costs which were incurred in the design was A$100 (500 s of CPU time on a CYBER 171 computer). Only one cycle was required for convergence.

Table 4.1 Optimum design solutions.

	Z_p for rigid–plastic design (cm^3)	Z_p for strong column design (cm^3)
B6	1004	1022
B5	2191	2249
B4	3042	4432
B3	4008	3995
B2	4762	4286
B1	5381	4173
C6	981	1498
C5	2027	2564
C4	3271	3684
C3	4343	5138*
C2	5246*	6677*
C1	6068*	8144*

*Note: this example is provided for demonstration only, and no allowance has been made to restrict the member sizes to the existing range of universal sections.

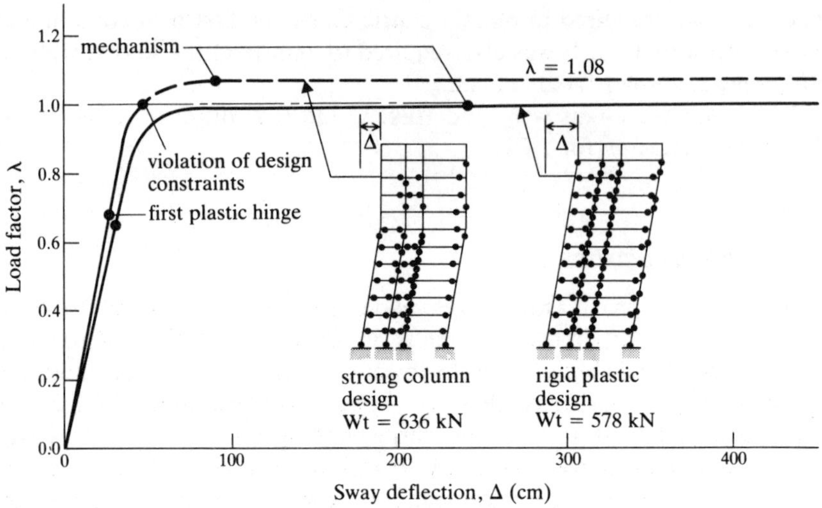

Figure 4.6 Load–deflection diagrams for load condition LC2.

4.2.2 Strong column design

Because the development of plastic hinges in columns tends to lead to premature buckling, it is often thought advantageous to confine locations of plastic hinges to the beams of the frame. A particularly desirable mechanism condition for earthquake loading is the one shown in Figure 4.7a. Plastic hinges form in the beams and bases of the ground floor columns, and the mechanism involves the entire structure.

A second so-called 'strong column' design was therefore carried out, and was characterized by the following criteria:

(a) LC1 (vertical loads) – frame to remain elastic as in the previous design.
(b) LC2 (vertical + earthquake loads) – frame should not collapse, and plastic hinges to be restricted to the locations shown in Figure 4.7a.

The results are also shown in Table 4.1, and the corresponding pattern of plastic hinge formation for LC2 at the prescribed unit load factor is shown in Figure 4.7b. As illustrated in the figure, design criterion (b) was satisfied, but the number of plastic hinges which formed was far less than is shown in Figure 4.7a. In particular, only one of four plastic hinges formed in the four base columns. This is perhaps not too surprising a result considering the extensive moment redistribution required to develop those

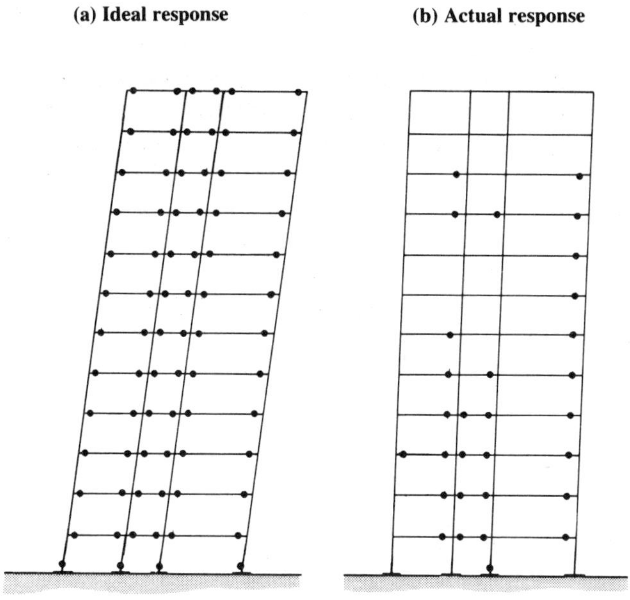

Figure 4.7 Strong column design – response to load condition LC2 at $\lambda = 1.0$.

four hinges. An extensive redistribution is required because of the effects of axial force on moment capacity, and because of the large variations in the axial forces in the four columns under the combined vertical and horizontal loading.

As indicated by the upper curve in Figure 4.6, the four hinges eventually formed in the base columns, but only after the load factor was increased above the design limit of $\lambda = 1.0$, and plastic hinges were allowed to develop anywhere in the frame. As shown in the figure, failure eventually occurred at a load factor $\lambda = 1.08$, and a sway deflection $\Delta = 91$ cm.

It is informative to compare the strong column design with the previous rigid–plastic solution. Although there was an improvement in load capacity, there was also an increase in frame weight. The collapse mechanism was reached at a far smaller deflection, and was associated with four fewer storeys. The strong column design was therefore characterized by a reduced capacity to absorb the energy of the earthquake. It is clear from this example that a satisfactory design solution is not necessarily obtained simply by preventing plastic hinges from forming in columns. Another method of avoiding premature buckling should therefore be sought.

Such a method has been developed by one of the authors (Zeman 1975), and involves considering buckling effects in the design procedure itself. Even in the case of a structure capable of the ideal response illustrated in

Figure 4.7a, the large deflections incurred are associated with P–Δ buckling effects which need to be considered. They lead to reductions in both load carrying and energy absorbing capacities. The method used is an extension of the one described in this monograph, and involves modifying the design equations in Chapter 3 to include second order theory. In each design cycle, response and response sensitivities are calculated on the basis of stiffness which is not only a function of the section sizes, material properties and geometry of the structure, but also of the axial forces in its members. The result is a design procedure which incorporates buckling effects with only marginal increases in complexity and computing cost. This procedure is beyond the scope of this monograph, and will be the subject of a subsequent publication.

References

Zeman, P. 1975. *Optimum elastic–plastic design of framed structures, Vols I and II. Ph.D.* thesis, University of Sydney, Australia.

Index

The boldface entries, e.g. **19**, indicate that the treatment of the subject concerned is given on that *and* the following pages.

analysis 1, **7**
  elastic 7–10
  elastic–plastic 9–20
  load history 4, **10**, 45, 70, 72–3, 77
analysis–design analogy 20–3, 43, 70
axial force 3, 51, 53, 75, 80

bending moment 3, 7, 8, 11, 53, 65
  see also moment capacity
buckling 83
  see also second-order theory

cable-stayed bridge 46, 50
Choleski method 56, 61
Colonnetti 2, 4
compatibility 2, 54, 61
computer 1, 75
concrete 3
constraints 44
  deflection 2, 32, 36, 78
  ductility 32, 33, 68, 72
  move limit 52, 79
  serviceability 32, 69, 71
  yield 16, 23, 28, 32, 40, 51–2, 65–8
continuous beam example **6**
contour *see* design
convergence 4, 79
CPU time 78, 81
crane 75
criteria 1, 2, 4, 6
  see also constraint
cross sectional area *see* section property
CYBER 78, 81

Davies 3, 4
deflection 2, 3, 4, 8, 69, 78
  see also constraint
deformation 3, 53, 54, 61
  see also deflection
design 1, 20
  contour 24, **33**
  cycle 4, 27–8, 32, 44, 49, 53, 78, 81
  elastic 1, 3, 4, 22–32
  plastic 32–43
  proportional 35, 43
  rigid–plastic 1, 2, 3, 35, 77, 81
  space 24, 33, 36
discrete list of sections 48

ductility 2, 68
  see also constraint

earthquake 80
elastic *see* analysis and design
elastic domain 25
elastic–plastic *see* analysis
equilibrium 8, 54, 61

factored loads 6–7
feasible 24
flowchart 45
formulation 2, **27**, **39**, 44, **53**, **64**
frame 2, 75, 79

Gass 2, 4
Gaussian elimination 18
geometry *see* variable

haunch 75
Heyman 1, 2, 4
hinge
  freezing 73
  plastic 2, 10, 17, 32, 72, 78
  rotation approach 2, 10
  see also variable
Horne 2, 4
Hung 3, 4

instability 2
  see also buckling, $P-\Delta$ effects

joint rotation 7

Krishnamoorthy 3, 4

limit state 1
linear program 2, 3, 10, 14–22, 41, 51, 64
linearization 48
load capacity 8, 9, 20
load condition 4, 6, 75, 80
load-deflection diagram 20
load factor 75
load history *see* analysis
load parameter 6
  see also variable

Macchi 2, 4

Maier 3, 5
mechanism 2, 10, 20, 35, 82, 83
member size 3, **46**
see also variable, section property
minimum weight 2, 4, 64
see also objective
moment area method 9, 11
moment capacity 6, 7, 9
see also variable
Morris 2, 4
move limit *see* constraint
multistorey frame example 79–83
Munro 3, 5

non-linear 4, 79

objective 4, 16, 24, 32, 45, 50
operations research 2
optimization 2, 16, 24, 27, 40, 42, 45, 69, **70**

$P-\Delta$ effect 2, 3, 83
see also second-order theory
permanent deformation 8
plastic *see* hinge, analysis, design
portal frame example 75–9
prestress *see* variable

reinforcement 3
Reinschmidt 3, 5
response sensitivity 53, 59–64
rigid arm 75
rigid–plastic 1
see also analysis, design

second moment of area *see* section property
second-order theory 2
see also $P-\Delta$ effect
section modulus *see* section property
section property 47–9
cross-sectional area 23, 47, 66
second moment of area 7, 23, 47
section modulus 48, 66
section size 2, 4, 7, 75, 79
see also section property, member
seismic 79
serviceability *see* constraint
shape factor 8, 21
slack *see* variable
slope deflection 7
SODA 3, 75
statics matrix 54, 57, 61, 62
steel 2, 3
stiffness approach 1
stiffness matrix
frame 55, 61
member 54
stress resultant 53, 61
strong-column design 82
superposition 12, 54
sway 81

Taylor-expansion 53, 67, 68
Toakley 2, 5
ultimate load 2, 4, 32, 46
universal section 47, 75, 79

variable 4, 45–50, 65
geometry 4, 50, 62
hinge rotation 3, 12, 16, 17, 32, 39, 51, 60, 64
load factor 70
member size 4, 46–9, 59, 69
moment capacity 21, 32, 33
prestress 4, 51
slack 17, 30, 42, 71
transformation 72, 73

wind 75
Wood 2, 5
working load 2, 4, 46, 78

yielding 3, 7, 8, 9
see also constraint
Young's modulus 7

Zeman 3, 5, 54, 83